The
Constraints
Management
Handbook

The St. Lucie Press/APICS Series on Constraints Management

Series Advisors

Dr. James F. Cox, III
University of Georgia
Athens, Georgia

Thomas B. McMullen, Jr.
McMullen Associates
Weston, Massachusetts

Titles in the Series

Introduction to the Theory of Constraints
by Thomas B. McMullen, Jr.

Securing the Future: Strategies for Exponential Growth Using the Theory of Constraints
by Gerald I. Kendall

Project Management in the Fast Lane: Applying the Theory of Constraints
by Robert C. Newbold

The Constraints Management Handbook
by James F. Cox, III and Michael S. Spencer

The
Constraints
Management
Handbook

Foreword by
Eliyahu M. Goldratt

James F. Cox, III, Ph.D., CFPIM, Jonah Instructor
Michael S. Spencer, Ph.D., CFPIM, Jonah Instructor

The St. Lucie Press/APICS Series on Constraints Management

S^t_L

St. Lucie Press
Boca Raton, Florida

Library of Congress Cataloging-in-Publication Data

Catalog information may be obtained from the Library of Congress

© 1998 by CRC Press LLC
Lewis Publishers is an imprint of CRC Press LLC

No claim to original U.S. Government works
International Standard Book Number 1-57444-060-8
Printed in the United States of America 1 2 3 4 5 6 7 8 9 0
Printed on acid-free paper

S^t_L

St. Lucie Press
2000 Corporate Blvd., N.W.
Boca Raton, FL 33431-9868

APICS
500 West Annandale Road
Falls Church, Virginia 22046-4274

TABLE OF CONTENTS

FOREWORD

During the eighties the major fear of middle managers was that their top management would decide to embark on another "improvement program." No wonder, new movements were popping up one after another and top managment, facing intensified competition, were imposing them on their organizations. Jumping from one program to another became such a widespread phenomena that in most companies the cynical phrase, "the program of the year," did not need any explanation.

The eighties started with the MRP II crusade, or as many interpreted it, computerization. Then the quality movement—TQM—became popular almost overnight. Before people had time to understand what TQM really means, Just-In-Time (JIT) appeared on center stage, the movement that turned the perception of inventory from an asset into a liability. In 1984, the book *The Goal* started to become popular, and the *Theory of Constraints* (TOC) was introduced. It made perfect common sense but at the time it seemed most controversial. Toward the end of that important decade, human-related movements started to be advocated, empowerment, and team work.

Now we are in the second half of the nineties, and we see that all those movements did pass the test of reality, with flying colors. No one doubts that computers are the obvious tool to deal with the logistical data. The perception of what is acceptable quality has changed once and for all. There is no need anymore to emphasis the importance of reducing inventory, it's common place. The distinction between bottlenecks and non-bottlenecks is spread world-wide. And it is hard to find a manager who argues that empowerment and teamwork are not mandatory for success.

These movements were not simply, "the program of the year," they are here to stay. Moreover, a new realization is now emerging, the realization that we cannot choose one and ignore the others, they are all essential pieces of the same puzzle, of the new management framework.

It shouldn't come as a surprise. All these movements had many things in common. The most obvious one is that they all independently advocated embarking in the same direction, they were all zealous about directing organizations into a "process of on-going improvement."

The less obvious common thread, but probably more important, is that they all also advocated a "system approach." This is not a coincidence, it is because in every case it is the foundation of the new knowledge each movement brought. What is a "system approach?" I interpret it as a warning against a common practice: concentrating on a local optima (in place or time) and, by that, jeopardizing the performance of the system as a whole.

It is easy to see it clearly in the tangible techniques of these movements. For example, the KANBAN technique of JIT. What is it if not a technique that is built to prevent the local optima of a worker producing all the time, producing even when down-stream operations do not now need the results of his/her work?

The tangible technique from TQM is known as Statistical-Process-Control (SPC). The main message of SPC is that as long as the deviation of a process is within the statistical boundaries, the biggest mistake a person can make is to re-align the process. It is a warning against a local optima in time; at this instant you may think that you have brought the process to be closer to the average but actually you enlarged the deviations of the system.

It is also obvious that empowerment and teamwork were not widely used before because of the tendency to strive for local optima in time and place.

As for TOC, its motto is, "Local optima do not add up to the optimum of the system as a whole." The key to know what to do locally is the realization of the role the system constraints are playing.

We have the pieces, at least many of them. Now the challenge is to clearly construct this new puzzle; the new management framework. We have to find out exactly how the pieces fit, to sort out the confusion in the places they overlap, to realize where there are still blind spots. And people like Jim Cox and Michael Spencer are rising to this challenge.

The Constraint Management Handbook is an excellent attempt to unify the pieces in production: sorting through the fog of classification, the overlap between MRP, JIT, and TOC, and clarifying the implications on one of the most important aspects—the measurements.

On these topics Cox and Spencer have done a comprehensive work. But they don't stop there. They also give a firm base for others to continue in this effort. They introduce the reader to the power of uniting logic in a systematic way—the thinking processes—so that more people will help in the effort to turn the field of management into a hard science.

Eliyahu M. Goldratt
Kfar-Sava, Israel
July 1997

PREFACE

Our interest in Theory of Constraints (TOC) dates back to Goldratt's early presentations on performance measures and production at the APICS International Conferences in the early 1980s. What was originally introduced as a software package, OPT, and a set of rules of scheduling lead to challenging the basis of planning and control in manufacturing. Many of the OPT concepts challenged not only cost and managerial accounting but many of the traditional business concepts. In an early survey of OPT users, several indicated that the software was excellent but its success would be limited unless businesses changed their performance measurement systems. *The Goal,* Goldratt's best-seller novel, laid the foundation for changing the way managers' viewed manufacturing. Many companies were quite successful based on implementing the constraints management (CM) concepts presented in the novel. Based on these results, in 1986 Goldratt and Fox founded the Avraham Y. Goldratt Institute, an educational institute for the development and dissemination of knowledge in TOC. The Institute has developed TOC tools to improve production, distribution, and supply chain management, project management, marketing, strategy, and thinking skills. The thinking skills include day-to-day management tools and more sophisticated tools to address massive problems (no matter what the field). From its early beginning as a forward finite scheduler to today's views of TOC as a business philosophy and as common sense, a continuous improvement process is present—focus on the system's goal, determine how to measure improvement (movement towards the goal), identify the blockage to achieving the goal (commonly referred to as the constraint), manage (exploit or eliminate) the constraint, subordinate all else to the constraint, and continually reevaluate the system with this process.

The CM concepts, whether measures, scheduling, buffering, thinking tools, etc. are extremely robust. They represent a totally different way of viewing, identifying, and solving problems—any problems. This book, while it focuses on the manufacturing logistics concepts, presents a flavor of the measurement issues and the thinking tools, Section I contains five chapters. Chapter 1 provides a simple explanation of constraints management from a production perspective. Chapter 2 provides a new conceptual framework for manufacturing. Chapter 3 introduces the five-step focusing process of continuous improvement. Chapter 4 describes drum-buffer-rope scheduling and control. Chapter 5 introduces V-A-T analysis and the production control points. Planning and controlling these few points simplifies managing any logistics system.

Section II of the book is comprised of four chapters. In Chapter 6 an overview of the V, A, and T plant cases is provided. Chapters 7, 8, and 9 provide detailed case descriptions of the company, production planning and control methods, key problem areas and an analysis of the company for a V plant (Valmont/ALS), T plant (Trane) and A and V plants (Stanley). These cases are rich in the details of using CM.

Section III is comprised of two chapters. Chapter 10 provides an overview of the performance measurement problem and the CM approach to managerial decision making. The CM performance measurement system is aligned with economic theory on the concept of fixed and variable costs. More importantly, CM measures have an outward focus linking resource capability to market opportunities in achieving the system's goal. Chapter 11 presents an overview of the thinking tools. These tools represent a paradigm shift in thinking. They provide the basis for scientific thinking—applied logic. Many CM users remark that CM is nothing more than common sense. The problem has been that most managers do not apply common sense to their business problems. As Goldratt states "people apply common practice."

This book provides a snapshot at a rapidly changing field (business) and at a rapidly expanding set of tools and applications. Recognize that mastering CM is a journey, not a destination. Our advice to everyone is to read Goldratt's materials on CM and the materials in this series. Many of the books are dedicated to a single topic and provide more depth than the overview provided here. This is a set of skills that you should master—it will significantly improve your business, professional, and personal decision-making.

Goldratt, E.M. and J. Cox, *The Goal: Excellence in Manufacturing.* Croton-on-Hudson, NY: North River Press, 1984.

ACKNOWLEDGMENTS

Mike and I (Jim) would like to acknowledge several individuals. Without their help we would have been unable to complete this work. First, our families provided us continuing encouragement throughout the project. We appreciate their understanding and love. During the writing of this book, both Mike and I (Jim) had daughters get married. Also, we both had children go off to college. Both of our wives managed weddings, sending kids off to college, and the many other major life events while we worked on this project. In addition to his wife, Susan, Mike gives a special thanks to his children, Kathleen, Cynthis, Kelsey, Andrew, and Austin. In addition to his wife, Mary Ann, Jim gives a special thanks to his children, Laurie Ann and Jimmy.

Second, we appreciate the numerous managers and staffs in each of the firms studied. Their patience and guidance in our understanding their specific environments and problems were considerable. We also would like to thank Jim Wahlers for the use of his Constraints Management case (Velmont ALS).

Third, APICS Dictionary definitions are reprinted with the permission of APICS The Educational Society for Resource Management, Falls Church, VA, APICS Dictionary, 8th edition, 1995. APICS is an excellent resource for educational meterials on constraints management and a leader in research and programs in this area with the efforts of the Educational and Research Foundation and the APICS constraints management SIG.

Fourth, Dr. Eliyahu M. Goldratt, the leading force in developing the Constraints Management body of knowledge, has continually shared his ideas and educational materials freely with academics around the world. We have used his ideas and educational meterials in our teaching for over a decade. Eli is an educator in the truest sense of the word. His Cocratic approach to teaching and to developing knowledge in each student is rare and through this method he creates individual ownership in discovery of the knowledge time and again. His teaching style and thinking processes provide the student the rare and valuable skills necessary to logically think through complex situations to get at the heart of a problem. While this text is about constraints management applied to manufacturing, Goldratt's concepts of being goal-oriented, having proper measures, and focusing on what is blocking achievement of the goal apply to all environments-for-profit, not-for-profit, government, service, manufacturing, personal, professional, and family organizations. Once mastered, the thinking tools will provide the individual effective communication skills, problem-solving skills, decision-making skills, and leadership skills.

Fifth, numerous professionals contributed to our understanding of constraints management. Bob Fox was a partner with Eli for several years. Bob also helped develop many TOC concepts and taught many TOC courses. He has always been supportive of our use of his knowledge and materials in our teaching. Bob is a true business professional and is always concerned about the students' understanding of the body of knowledge. Bob and Eli provided me (Jim) the opportunity to attend numerous TOC workshops over the years and also to teach practitioners and academics several different TOC courses over the years. These teaching efforts increased my TOC knowledge immensely. Many professionals in the Avraham Y. Goldratt Institute have contributed to our understanding of the constraints management body of knowledge. Dale Houlse, Donn Novotny, Dee Jacob, Alex Klarman, Alex Mashar, and Oded Cohen stand out as true leaders in Theory of Constraints. They are both excellent researchers and teachers. John Covington, Eli Schragenheim, Lisa Sheinkopf, Tom McMullen, Kevin Fox, Keith Krantz, Mike Umbel, Charlene Spoede, and other consultants and academics have been invaluable in expanding our understanding of constraints management. Each has been an early pioneer in extending this body of knowlege in his/ her environment.

Sixth, over the years, several of the doctoral students in Operations Management at the University of Georgia have increased our understanding of Constraints Management through their dissertation research. Mike Spencer was a friend and on the Board of Directors of APICS with me (Jim) for a few years. His case research provides the basis for Chapters 8 through10. Mike brought with him a detailed knowledge of JIT and MRP. We had many long discussions on the merits of MRP, JIT, and TOC. Jim Wahlers conducted his research on job-shop performance measures. Dick Moore conducted his research on large-scale inventory systems. He accepted a teaching assignment at the Air Force Institute of Technology. Dick was instrumental in converting the U.S. Air Force to using Theory of Constraints. These TOC concepts provide the foundation of the lean logistics system of the U.S. Air Force. Dick is now a partner with Bob Fox in the Theory of Constraints Center and is teaching industry how to effectively use these tools. Marsha Kwolek, another Air Force officer, was our first Ph.D. student to utilize the Current Reality Tree in case study research in analyzing Air Force Depot maintenance. Archie Lockamy and I wrote a book on performance measurement based on research from his disseration. Rex Draman researched the use of TOC as the basis of strategic planning. He spent many days with executives of a small firm using the CRT, FRT, PRT, and TRT tools identifying major clients' core problems, the firm's core problems and determining business strategies to support win–win solutions. Paul Pittman completed the first dissertation on using Goldratt's constraints management concepts in project management. Ed Walker is extending these concepts to the multi-project environment. Jeff Scott is studing the use of TOC measures and strategic placement of inspection and improvements in total quality management efforts in VAT environments. Lynn Boyd is currently researching the linkages between accounting systems and production planning and control systems for effective decision making. He is analyzing the use of direct costing, throughput accounting, absorption accounting, and activity-based costing with reorder point/economic order quantity inventory systems, MRP, JIT, and TOC planning and control systems on the major decisions in an organization through survey, case study, and simulation methods.

Seventh, John Blackstone was an early proponent of Theory of Constraints in teaching and research. He also lead several TOC dissertation efforts at the University of Georgia. Under John's directions, Ed Wray is extending the strategic buffering and buffer sizing concepts in projects. Brian Atwater and Jim Putt investigated buffering in lines. Stan Gardner investigated drum-buffer-rope scheduling in a job shop environment. Satya Chakravorty investigated TOC in supply chains. Dan Guide investigated drum-buffer-rope scheduling and TOC in remanufacturing. We have learned a lot from these research efforts. John has continually been a source of knowledge in constraints management and a sounding board for ideas.

Last, we would like to thank hundreds of APICS members across the country that attend our APICS presentations and workshops. It is through their questions and their sharing of their thoughts and problems that we have a better understanding of the commonalities and uniqueness of organizations today.

Jim Cox
Mike Spencer

ABOUT THE AUTHORS

JAMES F. COX III

James F. Cox III, Ph.D., CFPIM, is the Robert O. Arnold Professor of Business in the Terry College of Business at the University of Georgia. He received his Ph.D. in Engineering Management from Clemson University. Prior to his academic career, he held positions in construction engineering, industrial engineering, and production planning and control.

Dr. Cox is a certified Jonah and Jonah Instructor in the Theory of Constraints by the Avraham Y. Goldratt Institute. In addition to teaching Jonah workshops to the Air Force in support of its Lean Logistics efforts, he has conducted numerous academic and practitioner Theory of Constraints workshops and programs on performance measurement, production, supply chains, management skills, and the thinking processes.

Dr. Cox's research has centered on the Theory of Constraints for the past decade. He has published research on performance measurements, logistics, and the thinking processes. *Reengineering Performance Measurement with Lockamy* was one of the first books on constraints management performance measures. He has published over 70 articles and 5 books. He is a co-editor of the APICS Dictionary, seventh and eighth editions, and an invited contributor on *Constraints Management to the Production and Inventory Management Handbook*.

Dr. Cox has been a member of APICS for 20 years, holding chapter, regional, and national offices. He has served on the APICS Board of Directors for 4 years with 2 years as Vice President of Education — Research and has served on the APICS Educational and Research Foundation Board of

Director for 6 years with 2 years as President. He is a charter member of the Constraint Management SIG and the Society of Systems Improvement. He has spoken at over 50 APICS and other professional-organization chapter meetings, several regional seminars, and several international conferences on Theory of Constraints.

MICHAEL S. SPENCER

Michael S. Spencer is currently an Associate Professor of Production and Operations Management at the University of Northern Iowa in Cedar Falls. He received a Ph.D. in Operations Management at the University of Georgia in 1992 and holds an M.A. in Economics (1981), and an M.B.A. (1974) and a B.S. in Economics (1971). His research interests are in the management of production systems in repetitive service and manufacturing throughout the logistics function.

Prior to receiving the Ph.D., Dr. Spencer had been employed for 11 years in various materials management positions at the John Deere Engine Division. Among the assignments were his participation in the design and implementation of both MRP and JIT systems. Dr. Spencer is certified as a fellow level of the American Production and Inventory Control Society (APICS) and served on the APICS board of directors from 1987 to 1990. He is currently serving as Vice President of the Educational and Research Foundation of APICS.

ABOUT APICS

APICS, The Educational Society for Resource Management, is an international, not-for-profit organization offering a full range of programs and materials focusing on individual and organizational education, standards of excellence, and integrated resource management topics. These resources, developed under the direction of integrated resource management experts, are available at local, regional, and national levels. Since 1957, hundreds of thousands of professionals have relied on APICS as a source for educational products and services.

- **APICS Certification Programs**—APICS offers two internationally recognized certification programs, Certified in Production and Inventory Management (CPIM) and Certified in Integrated Resource Management (CIRM), known around the world as standards of professional competence in business and manufacturing.
- *APICS Educational Materials Catalog*—This catalog contains books, courseware, proceedings, reprints, training materials, and videos developed by industry experts and available to members at a discount.
- *APICS—The Performance Advantage*—This monthly, four-color magazine addresses the educational and resource management needs of manufacturing professionals.
- *APICS Business Outlook Index*—Designed to take economic analysis a step beyond current surveys, the index is a monthly manufacturing-based survey report based on confidential production, sales, and inventory data from APICS-related companies.
- **Chapters**—APICS' more than 270 chapters provide leadership, learning, and networking opportunities at the local level.

- **Educational Opportunities**—Held around the country, APICS' International Conference and Exhibition, workshops, and symposia offer you numerous opportunities to learn from your peers and management experts.
- **Employment Referral Program**—A cost-effective way to reach a targeted network of resource management professionals, this program pairs qualified job candidates with interested companies.
- **SIGs**—These member groups develop specialized educational programs and resources for seven specific industry and interest areas.
- **Web Site**—The APICS web site at http://www.apics.org enables you to explore the wide range of information available on APICS' membership, certification, and educational offerings.
- **Member Services**—Members enjoy a dedicated inquiry service, insurance, a retirement plan, and more.

For more information on APICS programs, services, or membership, call APICS Customer Service at (800) 444-2742 or (703) 237-8344 or visit http://www.apics.org on the World Wide Web.

CONSTRAINTS MANAGEMENT LOGISTICS BASICS

WHAT IS CONSTRAINTS MANAGEMENT?

INTRODUCTION

What would happen in your company if someone had a breakthrough idea for managing the business? Consider the following scenarios and ask what the response of your company's management would be to these improvement efforts.

Scenario 1—A custom cabinetmaker routinely takes five to six weeks to make a custom-designed set of kitchen cabinets for a new home. A feature article was written in a leading practitioner journal when one company was able to reduce its lead time from six weeks to ten days using just-in-time (JIT) techniques. When this same company reduced its lead time to two days using constraints management (CM) techniques, it became a competitive advantage noone in the company wanted to discuss with outsiders.

Scenario 2—The electronics division of a leading automobile manufacturer reduced its average manufacturing lead time for electronic harnesses from a few months to less than two weeks by using JIT techniques. After applying CM techniques to these same processes, the lead time for 95% of all harnesses was reduced to less than two shifts.

Scenario 3—A producer of lickless postage stamps could produce two million booklets a week using a continuous manufacturing process (four shifts for seven days a week). Using CM, this same facility with an 8% increase in capital equipment is now producing 25 million booklets a week on a three-shift five-day-a-week time frame.

Scenario 4—A continuous-process chemical plant increased output 10% immediately by inserting buffers before and after the constraint resource. The buffers act as an information system to focus improvement activities on equipment maintenance. Management predicts that when its focused maintenance projects are completed, the output will be 50% over the initial process output. When asked how the cost of the CM improvement projects compared to building a plant to provide the 50% improvement in output, management responded, "We've never looked at it like that—We estimate that the CM effort will cost between $100,000 to $200,000 compared to a $30 or $40 million plant expansion based on our traditional management practices."

Scenario 5—The president of the Institute of Managerial Accounting's Boston chapter (a retired group vice-president of Texas Instruments) announced to an audience attending the play *Uncommon Sense* (written by Eli Goldratt) at an Institute of Managerial Accounting chapter that "Texas Instruments improved existing operations using CM in the early 1990s to the extent that TI deferred over $1.2 billion in planned capital investments in two new plants."

One question almost always arises concerning CM: Where are the success stories? Many companies are using CM concepts without publicizing it. It is hard these days to find a production manager who has not read *The Goal* or to walk through a major airport without seeing a copy of *The Goal* tucked under someone's arm. The book has sold over two million copies and its CM concepts have been implemented successfully (and unfortunately unsuccessfully) in countless plants. Many companies that have specifically used the CM concepts are reluctant to advertise their results and, thus, give away their competitive advantage.

Constraints management has been used by companies such as Avery Dennison, Bethlehem Steel, General Motors, and Procter & Gamble. The United States Air Force Logistics Command has adopted CM concepts to improve performance of aircraft repair depots, and the United States Navy has implemented concepts in its Transportation Corps. The successes are numerous and quite significant if we look for and ask the right questions of the right people.

Most organizations recognize such tremendous improvements in organizational performance that their managers prefer not to divulge the sources of the dramatic improvements for fear competitors will follow their path of improvements. Constraints management is truly a paradigm shift, as JIT and total quality management (TQM) were shifts from traditional management practice. Three major differences between JIT/TQM and CM are that CM recognizes (1) the pitfalls of using traditional measurements to identify and measure improvement, (2) the infeasibility of implementing JIT/TQM throughout production, and (3) the problems of not involving all functions (marketing, sales, engineering, etc.) in continuous improvement efforts.

We're getting ahead of ourselves. Let's briefly review the basics of production to ensure a common understanding. Goods are produced in a production system. The production system consists of a series of successive steps performed by different resources. To build a product, all of the steps or operations must be completed in the specified sequence. As we shall see, one resource, the constraint, limits overall system output. Traditional approaches to production management view each operation as an independent activity to be managed and measured at each resource. Thus, these approaches lose sight of the objective of the overall production system—to produce and sell goods and services.

Constraints management is a new approach to planning and controlling the production and sale of goods and services. It recognizes the powerful role of the constraint (the limiting resource) in determining the output of the entire production system. By understanding and mastering the CM concepts, managers can see immediate improvements in their organizations' results and through a focused continuous improvement approach can plan for meeting future needs as well.

WHAT'S WRONG WITH WHAT WE ARE DOING NOW?

Throughout most of history, the production of goods and services was labor intensive and was very simple. Someone who wanted a chair went outside, cut down a tree, cut the tree into workable lumber, and used woodworking tools to fashion the chair. To make a better chair, more time was spent on the activities of making it. With the start of the industrial revolution, production methods changed. Manufacturing workers produced products for purchase and consumption. Specialization of labor replaced the craftsman. Specialization created low-skill, low-cost production methods. Each worker became a specialist in one aspect of production. Today, few workers, especially managers, are able to make much of anything. Management became the art and science of getting things done through other people. On a larger scale, production became a specialized area of business, as did accounting, finance, and marketing.

The planning and control of the production process are a large part of management. Various approaches to improve production have been developed over the years. Some work well for a time and then fail. Others don't work at all. Those that seem to work for a time tend to fail at often unexpected moments—like when a shipment to a key customer is due. Most production managers are accustomed to fighting fires and expediting shipments and may assume that is simply "business as usual." This view need not be the case.

Breakthroughs in thinking occur. Sometimes someone makes a substantial contribution by explaining why things don't work as expected. Occasionally, someone creates a new approach that actually does work based on an understanding of why things go wrong. When we determine why things won't work and create new methods, *the art of management becomes a science.* As a science, production management can be studied in an objective manner, subjected to tests, and results reported to others. This scientific process takes time and, of course, the explanations, the tests, and the results build upon one another into an overall body of knowledge.

The process of studying, understanding, testing, and reporting results does not progress smoothly. If it did, education would be easy, which it is not. There are stops and starts, and time passes when no progress is made. Then, suddenly, and unpredictably, an insight occurs and the pace quickens. The insight might be sparked by the development of a new technology, as was the case with the computer and material requirements planning, or someone from a different field might simply bring a fresh perspective to the field, as is the case with constraints management. Eliyahu M. Goldratt, a Ph.D. physicist, is the driving force in developing the new management philosophy of Theory of Constraints (or constraints management).

The answer to the question posed by the title of this section—What's wrong with what we are doing now?—is both nothing and everything! Nothing is wrong with what we are doing now because we are able to expedite materials, schedule overtime, repair equipment, and through these extraordinary actions make shipments to customers.

Production management has succeeded in providing an ever-increasing array of goods and services to an ever-growing consumer population. We should not make light of this amazing accomplishment. Certainly, the former Eastern-bloc countries, the former Soviet Union, and the Third World countries are quite well aware of the U.S. accomplishments in manufacturing, so much so that they seek to emulate US manufacturing processes. The answer to our question—What's wrong with what we're doing now?—is also *everything is wrong,* because to accept nothing is wrong as the answer is to stop the entire scientific process.

Wickham Skinner,[1] a noted retired Harvard scholar, provides ten reasons why operations management is failing to produce significant results:

1. We've done a lot but it will take more time for it all to take effect.
2. Foreign competition has been getting better, too. We're just chasing and not catching up.
3. Top management still doesn't understand and really support manufacturing.
4. We are going much too slow in investing in automation.
5. Employees are still a big problem, both in mediocre enthusiasm and in skills and education.
6. The right kinds of managers in operations are lacking.
7. It's the other departments that are holding us back, i.e., accounting, finance, marketing, engineering, purchasing, personnel.
8. Alphabet soup! Too many uncoordinated, off-the-shelf programs and activities sold by consultants and professors.
9. The playing field isn't level. Our costs for health care, the "litigious society," environmental improvements, retirement and Social Security, and costly work rules make competing with many foreign nations virtually impossible.
10. As a society and culture, the United States is "post-industrial." People just don't want to work in industry anymore. We're trying to push a big rock uphill. History is against us.

Skinner's list of causes is impressive, but it focuses on the effects and not the causes of our dilemma. JIT, TQM, and CM have created a paradigm shift where the customers' expectations have changed significantly. Customers expect cheaper, better, faster, customized products and services. Management can no longer afford to treat the organization as separate, independent activities and functions. Management must learn to respond to customers' expectations in order to remain competitive and on a path of continuous improvement. In this handbook, we will focus primarily on the CM concepts to improve the production process. The reader should recognize that an understanding and mastery of CM measures and the Thinking Processes is essential for true continuous improvement of any organization.

What would happen if we stopped improving the management of the production process? Are our traditional management approaches the best that can be used on the shop floor or in the planning office? If we have learned anything over the past several years, it is that to stop is to actually retreat, as others (our competitors) will not stop. Advancement is relative. Certainly, change is difficult and often painful, but the alternative is worse. The Japanese created the first paradigm shifts in production management with their focus on achieving high quality and reducing lead time through TQM and JIT. In this chapter, we will discuss why the traditional ways of managing production fail—fail not in the sense of not producing goods and services, but fail in the sense of denying us the next step in the process of continuous improvement. We will also present an overview of the constraints management approach and demonstrate how it improves production management.

WHY TRADITIONAL PRODUCTION MANAGEMENT FAILS

Let's look at a very simple production example provided in Figure 1.1. Product C is assembled from two components, A and B. The components start as raw materials A and B, respectively, and each goes through three different operations at different work centers. Raw material A goes through operations (opn) 10, 20, and 30, while raw material B goes through operations 15, 25, and 35. Components cannot skip an operation and then go back, nor can they be produced at any other work center. Each operation requires a certain amount of production time at each work center, and because different physical operations are performed at different work centers, the times and rates vary from one operation to another. Operation 10 can be performed at a rate of five units per hour and uses raw material A, operation 20 can be performed at a rate of two units per hour, and operation 30 can be performed at a rate of five units per hour. Raw material B is used at a rate of ten units per hour at operation 15, its first step; operation 25, the second step, at a rate of four units per hour; and at the last step, operation 35, at a rate of five units per hour. Assembly of component A and component B into product C is rather simple and takes only three minutes, or a production rate of 20 units per hour. Now, let's take a short quiz.

What is the maximum output of product C? If your answer is two units per hour, you are correct. Anything else is wrong. Let's look at Figure 1.1 again. As many as 20 products per hour could be assembled, since it takes only three minutes to assemble each one. However, the two components that make up the assembly take much longer to manufacture. Component B's first operation, operation 15, is produced in six minutes, for a rate

---

EXAMPLE

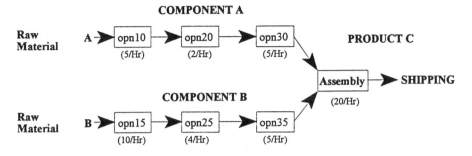

COMPONENT A

( ) = Quantity that can be produced per hour

QUIZ

- What is the maximum output of product C?   Why?

- What happens if operation (opn) 25 increases its output to 10/hr through increased efficiency?   Why?

- What happens to the output if more than 2/hour are released into operations 10 and 15?   Why?

- What happens if operation 10 breaks down for 4 hours?   Why?

- Is the result different from a breakdown of operation 30?   Why?

Figure 1.1   A Simple Production Example and Quiz

of ten units per hour. But all of the other operations take longer. Operations 10 and 30 of component A and operation 35 of component B each take 12 minutes to produce one unit. We seem to be closing in on the answer—and the heart of constraints management.

The slowest operation in this case, operation 20, is used to make component A and determines the output of the entire system. Operation 20 takes 30 minutes for each unit; since each product C takes both an A and a B component, no more than two units per hour can be manufactured even though many more can be assembled. The output of the slowest work center determines the output of the entire production system.

This concept will be refined later, but if you understand this concept, you can see why it is critical to manage the constraint operation differently than all others. Let's look at another question which will highlight the differences between CM and traditional management.

What happens if operation 25 increases its output to 10 units per hour by increased work center efficiency? The answer, given our previous discussion, is relatively clear. There is no change in the output of product C. A natural inclination may be to start moving people around in our imaginary system or to vary the times the operations are working. You might suggest something like, "We should work operation 20 on three shifts and operation 25 on two shifts..." But that is a more complex problem than we are discussing here, and, besides, CM points out another problem discussed later.

Even though most managers understand that the output of product C is unchanged, the implications are enormous. What would happen in a typical factory if a bright industrial engineer, computer program analyst, or quality circle work team devised a way of increasing the output of operation 25 by 250%, from one unit produced every 15 minutes to one unit every six minutes for a cost of $1000? Most management would leap for joy. A cost-benefit analysis would surely demonstrate a very high return on investment based on, if nothing else, the labor and overhead reduction. But what did we just find out about the output of product C? The matter is made even worse if we were to think of this system as separate departments rather than operations. Who would dispute the value of increasing department 25's productivity by 250% for only $1000? Yet, CM indicates that there is no real savings, and $1000 is being spent for nothing but idle time. A few more questions will help illuminate the surprising insights of CM.

What happens to the output of product C if more than two units per hour of raw materials A and B is released into operations 10 and 15? Again, based on our previous discussion, nothing happens. Well, not exactly. Nothing happens to the overall output of product C, but there would be a buildup of inventory between operations 10 and 20 and between operation 35 and the assembly area, and the implications are again enormous. If more material than two units per hour is released into operation 10, a buildup of material will occur because operation 20 cannot produce more than two units per hour. Therefore, operation 10's productivity is limited to only 40% of its potential of five units per hour. Similarly, in operation 30, the operation(s) after the constraint operation has capacity to produce five units per hour but cannot. It will produce two units per hour.

Look at the effect of the constraint on the production of component B. Assume that a single machining department, somewhere else in the factory, produces component B on three separate machines run by three different workers. The schedule calls for making 100 product Cs so 100 component Bs can be made. Raw materials B is released into operation 15 at a rate of ten per hour. This release policy is followed in order to maximize productivity. Isn't that good? The material builds up in front of operation 25 because it can make only four Bs in an hour. The supervisor notices the buildup and operations 25 and 35 each make a setup. Depending on the schedule, overtime may even be scheduled on both machines to finish the component on time. Productivity looks great, except for operation 15, which has run out of raw material. What would most departmental supervisors do in this situation? They might demand the release of more raw material B into the department to keep operation 15 working. But will this action do anything to produce more component As at operation 20, the constraint? If there is no more output at operation 20 than two units per hour, what is the real effect of the supervisor's actions in the department making component B? Work center productivity looks good, but at what cost, as shipments of product C are the same.

What happens if operation 10 breaks down for four hours? Machines break down, and workers are sometimes absent. What happens? The answer is that it depends. If operation 10 is lost for four hours and no inventory is between operations 10 and 20, then operation 20 stops for four hours as well. But the results are actually much worse. After four hours, operation 10 begins to produce component A. Operation 10 has the potential to make five component As per hour, but operation 20 does not. Since operation 20 is the constraint, the output of product C falls by eight units. At a minimum, overtime would be required at operation 20 to replace the four hours of lost production of component A. If a small buildup of inventory had been allowed for, however, the results would be different.

If an inventory of ten units, for example, is in place between operations 10 and 20, no output is lost and no overtime is required at the constraint. Operation 10 is down for four hours. As soon as operation 10 breaks down, the inventory between operations 10 and 20 begins to deplete. Four hours later, only two units of inventory are left, but operation 10 is now running again. Because operation 10 can produce at a rate of five units per hour, the inventory bank is quickly refilled. Operation 10 cannot continue to produce at a rate of five units per hour because the same condition discussed above will be encountered. Clearly, inventory positioned at the constraint provides benefits to the system. Compare that situation to what happens under JIT manufacturing where zero inventory is the goal. Let's examine one last situation.

What happens if operation 30 breaks down? Again, the answer is that it depends, but differently than before. Operation 30 has a production rate of five units per hour. If operation 30 breaks down, it can catch up as long as operation 20, the constraint, keeps producing. If enough physical space is provided between operation 20 and operation 30, a temporary bank of inventory will build until operation 30 can produce again. If space is limited, then operation 20, the constraint, will have to stop. Although operation 30 can produce at five units per hour, it is unable to do so because operation 20 also had to stop.

If freestanding equipment is used and ample space is provided, this problem does not materialize. But, again, what happens if component A is being manufactured in a cellular layout? Or in a transfer line? Or in any low-inventory layout? Does production engineering build a temporary inventory space immediately beyond the constraint operation in the layout? Or, as in most cases, is space for a temporary inventory bank actually engineered out of the production system, as occurs when JIT production methods are laid out?

Some of the commonly used traditional and JIT approaches to plan and control production appear inappropriate and counterproductive when we use constraints management.

WHAT IS THE CONSTRAINTS MANAGEMENT APPROACH TO PRODUCTION MANAGEMENT?

For several years, interest and confusion have existed over the use of the term Theory of Constraints (TOC), the term constraints management, the book *The Goal,*[2] and the software package called OPT. The first area of confusion concerns the differences among the terms. Are these terms the same, does one replace the others, or are they different? Early works concerning these production methods used the term, optimized production technology (OPT), while only recently has the term, CM been used to describe the application of TOC. The second area concerns the elements that comprise TOC. No clear-cut categorization of the elements of TOC exists to differentiate it from OPT. The third area of confusion is the use of some of the practical concepts presented in the book *The Goal,* especially the scheduling methods. Are these methods part of OPT or part of CM? How much knowledge of CM does one need to implement the methods discussed in *The Goal*? In order to clarify these terms, we provide a brief history.

Two references on the origin of OPT and the subsequent development of TOC were written by their developer, Dr. E.M. Goldratt: a 1988 article in the *International Journal of Production Research*[3] and the foreword to the book *Synchronous Manufacturing,*[4] published in 1990. Goldratt states, "The original name [OPT], established in 1979, was optimized production timetables. The main thrust was computerized scheduling, and the thoughtware was in its infancy. During the following years, the fundamental understanding grew by such leaps and bounds that in 1982, the name was officially changed to optimized production technology" (p. ix).[4] Goldratt also states, "I was one of the prime originators [of OPT] and...I'm now convinced that computerized shop floor scheduling is just a small issue in the much broader problem of successfully running a manufacturing company" (p. 444).[3] The OPT software was modified in 1980 to broaden its application to job shop environments with the development of the HALT module. "The HALT concept reduced inventory without jeopardizing sales but its implementation exposed the fact that most resources (machines as well as workers) in the plants which implemented the software had considerable excess capacity and very few resources could be used to 100% of their available capacity without causing inventory to be inflated....The recognition of the contradiction between balanced flow and balanced capacity in an environment which has statistical fluctuations and dependent resources started to become clarified....The OPT rules started to be formulated with the growing understanding that the software's superiority stemmed from not its algorithm, but mainly from these underlying concepts" (p. 448).[3]

Goldratt's book *The Goal* was first published in 1984 to present, in a Socratic style, the global principles of manufacturing as known at that time. The words optimized production technology or OPT do not appear in the book. Considerable attention is devoted to the importance of the bottleneck and to the conflicts between traditional cost performance measurement systems and the real goal of a manufacturing company. However, the actual scheduling methodology was emphasized less. For example, the terms drum-buffer-rope scheduling and buffer management were also not included in the book. Surprisingly, a number of companies implemented concepts found in the book without using the OPT software. "...[T]he results achieved were in many cases better than those that involved both education and software, and in all cases were achieved with much less investment and much less time" (p. 453).[3]

Refinements in the software continued, primarily driven by a deeper understanding of the underlying manufacturing concepts. "Only in 1985 did it become clear that inventory and time are not two separate protective mechanisms, but actually one....Moreover it started to become obvious that what is important is the impact of the schedule on plant performance and not the validity of the schedule as judged by the output of the computer....To counteract this problem Version 56 of OPT was released near the end of 1985" (p. 452).[3] Goldratt continues, "At this stage only a fraction of our efforts were geared towards shop floor scheduling and much more devoted to finding conceptual replacements for the cost procedures by the use of throughput, inventory and operating expenses procedures" (p. 452).[3]

Goldratt's second book, *The Race,*[5] co-authored with Robert Fox, clarified the changes made in the Version 56 OPT software. "Nevertheless the difficulties our clients had in adjusting to Version 56 prompted writing *The Race* in which we clarified our understanding of this scheduling technique. The 'Drum-Buffer-Rope' approach was formulated and explained in that book" (p. 452).[3] Elements of drum-buffer-rope are found in *The Goal,* but the methodology was not presented as a coherent system.

Based on his investigation into the success of those firms that had implemented concepts discussed in *The Goal,* Goldratt reached three conclusions. First, a need existed to provide a way for companies to turn the process described in *The Goal* into a process of continuous improvement. Second, the formulation of a theory for running an organization was required. Third was the recognition that "...in most plants the DRUM is a CCR [capacity constraining resource] that is not a bottleneck....The driving force should not be time but the exploitation of the constraint" (p. 454),[3] The OPT software as of 1988 did not address this thought, although Goldratt states, "According to my knowledge and judgement the OPT software as it exists today [1988] is currently the most powerful and successful software for shop floor scheduling" (p. 454).[3] A revised edition of *The Goal* was published in 1986 to incorporate the continuous improvement idea. This edition is familiar to most readers.

The use of the term OPT, therefore, should be limited to the description of the *software* as it is currently available. The software provides a finite scheduling methodology that is based on the maximization of production through the bottleneck resource. The software contains elements of drum-buffer-rope but not solidified as an identifiable system. It does not contain the TOC performance measurement system, thoughtware, or other underlying concepts of how an organization operates, although the OPT software executes a schedule assuming some of these concepts exist in the production environment.

The emergence of the underlying concepts of how an organization operates drew increased attention. Goldratt devoted his efforts to the development of a general theory for running an organization rather than improvements in the scheduling methodology. "At the beginning of 1987, I adopted the current name—theory of constraints. A better understanding of the psychology caused me to make a shift from emphasizing rules/principles to a focusing-iterative process. Moreover, the significant ramifications that this process has for such areas as accounting, distribution, marketing, and product design almost forced that choice of words" (p. x).[4]

Beginning in October 1987, a series of monographs was published, entitled *The Theory of Constraints Journal,*[6] that described the various components of the general theory, or thoughtware. In 1990, Goldratt published a book entitled *What is this Thing Called Theory of Constraints and How Should It Be Implemented?*[7] that includes concepts developed in the various monographs. Also in 1990, another book, *The Haystack Syndrome,*[8] was published to codify and discuss the logistics components and performance measurements components (including the well known PQ exercise) of the general theory of the organization. Goldratt's TOC concepts were also presented and refined in the course work that was part of the workshops conducted by the Avraham Y. Goldratt Institute. In 1993, the first article to discuss the problem-solving or thinking process components of TOC was published.[9]

Theory of Constraints consists of the following components: (1) a logistics branch, with the scheduling methodologies of drum-buffer-rope and buffer management, and the V-A-T logical structures analysis (used for production line design and analysis as well as distribution system design and analysis); (2) a second branch consisting of the five-step focusing procedure, the performance measurement system (throughout, inventory, and operating expense), the application of throughput-dollar-days, and the application of product mix decisions; and (3) a third branch concerning the Problem-Solving/Thinking Processes consisting of effect-cause-effect (ECE) diagramming and its components (negative branch reservations, current reality tree, future reality tree, prerequisite tree, and transition tree), the ECE audit process, and the evaporating cloud methodology. Figure 1.2 depicts a schematic of TOC. Some general definitions of TOC are provided in Figure 1.3.

The first branch, logistics, consists of those elements of TOC which have the most visibility to operations managers. The scheduling methodology, drum-buffer-rope scheduling, provides detailed instructions at a few control points which manage the complete system based on the constraint

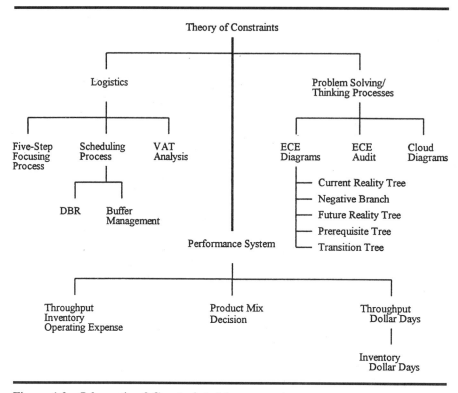

Figure 1.2   Schematic of Constraints Management

capabilities. V-A-T analysis is a classification of production processes that identifies the general product flow, the control points, and the strategic positioning of buffers. The term V-A-T originates from the shapes of the three diagrams that typically describe a production process for a product or product family. Those diagrams are based on the combination of the bill-of material structure and the component and assembly routings. Definitions of CM logistics terms are provided in Figure 1.4.

The second branch is compromised of the five focusing steps and the performance measurement system developed in *The Goal* to support the management of the constraint and to eliminate the conflicts with traditional performance measurement systems. The definitions of throughput, inventory, and operating expense have unique and specific meaning under CM which differ from the traditional definitions of these terms. The throughput-dollar-days and inventory-dollar-days are measurements to facilitate operating decisions. The

theory of constraints—A management philosophy developed by Dr. Eliyahu M. Goldratt that can be viewed as three separate but interrelated areas—logistics, performance measurement, and logical thinking. Logistics include drum-buffer-rope scheduling, buffer management, and VAT analysis. Performance measurement includes throughput, inventory and operating expense, and the five focusing steps. Thinking process tools are important in identifying the root problem (current reality tree), identifying and expanding win–win solutions (evaporating cloud and future reality tree), and developing implementation plans (prerequisite tree and transition tree) (*APICS Dictionary,* p. 85).[10]

constraint management—The practice of managing resources and organizations in accordance with theory of constraints (TOC) principles (*APICS Dictionary,* p 15).[10]

constraint—Any element or factor that prevents a system from achieving a higher level of performance with respect to its goal. Constraints can be physical, such as a machine center or lack of material, but they can also be managerial, such as a policy or procedure (*APICS Dictionary,* p. 15).[10]

Figure 1.3   Constraints Management Definitions. From Cox, J., Blackstone, J.H., and Spencer, M.S., *APICS Dictionary,* 8th ed., Falls Church VA: American Production and Inventory Society, 1995. With permission.

product structure—The sequence of operations that components follow during their manufacture into a product. A typical product structure would show raw material converted into fabricated components, components put together to make subassemblies, subassemblies going into assemblies, etc. (*APICS Dictionary,* p. 65).[10]

production network—The complete set of all work centers, processes, and inventory points, from raw materials sequentially to finished products and product families. It represents the logical system that provides the framework to attain the strategic objectives of the firm based on its resources and the products' volumes and processes. It provides the general sequential flow and capacity requirement relationships among raw materials, parts, resources, and product families (*APICS Dictionary,* pp. 65–66).[10]

drum-buffer-rope—The generalized technique used to manage resources to maximize throughput. The drum is the rate or pace of production set by the system's constraint. The buffers establish the protection against uncertainty so that the system can maximize throughput. The rope is a communication process from the constraint to the gating operation that checks or limits material released into the system to support the constraint (*APICS Dictionary,* p. 25).[10]

Figure 1.4   Constraints Management Logistics Terms. From Cox, J., Blackstone, J.H., and Spencer, M.S., *APICS Dictionary,* 8th ed., Falls Church VA: American Production and Inventory Society, 1995. With permission.

---

buffer—In theory of constraints, buffers can be time or material and support throughput and/or due date performance. Buffers can be maintained at the constraint, convergent points (with a constraint part), divergent points, and shipping points (*APICS Dictionary,* p. 10).[10]

buffer management—In the theory of constraints, a process in which all expedition in a shop is driven by what is scheduled to be in the buffers (constraint, shipping, and assembly buffers). By expediting this material into the buffers, the system helps avoid idleness at the constraint and missed customer due dates. In addition, the causes of items missing from the buffer are identified, and the frequency of occurrence is used to prioritize improvement activities (*APICS Dictionary,* p. 10).[10]

VAT analysis—A constraint management procedure for determining the general flow of parts and products from raw materials to finished products (logical product structure). A V logical structure starts with one or a few raw materials, and the product expands into a number of different products as it flows through its routings. The shape of an A logical structure is dominated by converging points. Many raw materials are fabricated and assembled into a few finished products. A T logical structure consists of numerous similar finished products assembled from common assemblies and sub assemblies. Once the general parts flow is determined, the system control points (gating operations, convergent points, divergent points, constraints, and shipping points) can be identified and managed (*APICS Dictionary,* p. 89).[10]

Figure 1.4 Constraints Management Logistics Terms (continued)

performance measurement system is useful in the product mix decision—the mix decision is based on the profit-per-constraint time period rather than the allocation of overheads in traditional accounting. Chapter 10 is devoted to the performance measurement issues and the CM performance measurement system. CM terms for performance measurement are provided in Figure 1.5.

The third branch of TOC is the problem-solving methodologies, called the Thinking Processes. This branch is the least researched and least visible to operations managers. The purpose of the problem-solving or thinking processes is to resolve three questions faced by managers in general: what to change, what to change to, and how to bring about the change. Definitions of CM Thinking Processes terms are provided in Figure 1.6.

The principal element of this branch is the ECE diagram. The methodology was originally based on the scientific method of postulating a hypothesized cause of an observed effect and then testing the causation by finding a confirming secondary effect. If a confirming secondary effect is found,

five focusing steps— In the theory of constraints, a process to continuously improve organizational profit by evaluating the production system and market mix to determine how to make the most profit using the system constraint. The steps consist of 1) identifying the constraint to the system, 2) deciding how to exploit the constraint to the system, 3) subordinating all non constraints to the constraint, 4) elevating the constraint to the system, 5) returning to step 1 if the constraint is broken in any previous step, while not allowing inertia to set in (*APICS Dictionary*, p. 31).[10]

throughput—In theory of constraints, the rate at which the system (firm) generates money through sales (*APICS Dictionary*, p. 85).[10]

inventory—In theory of constraints, inventory is defined as those items purchased for resale and includes finished goods, work in process and raw materials. Inventory is always valued at purchase price and includes no value-added costs, as opposed to the traditional cost accounting practice of adding direct labor and allocating overhead as work in process progresses through the production process (*APICS Dictionary*, p. 40).[10]

operating expense—In theory of constraints, the quantity of money spent by the firm to convert inventory into sales in a specific time period (*APICS Dictionary*, p. 55).[10]

Figure 1.5   Constraints Management Performance Measurement Terms. From Cox, J., Blackstone, J.H., and Spencer, M.S., *APICS Dictionary*, 8th ed., Falls Church VA: American Production and Inventory Society, 1995. With permission.

then evidence exists that the cause, as postulated, is true. The ECE diagram is a series of these postulated and confirmed relationships which lead to the uncovering of a primary cause or core problem which explains most of the observed undesirable effects. The primary use of the ECE diagram is to determine the root or core problem (often a policy, procedure, or other non-physical constraint) that may be causing several undesirable symptoms. The methodology is designed to force managers to uncover the true cause rather than to waste time solving symptoms only to have the real problem remain.

The second element, the ECE audit process, is the application of rules of logic (categories of legitimate reservation) to test and strengthen the relationships proposed in the ECE diagram. The audit process tests the ECE hypotheses and either confirms them as logical, as needing additional explanation, or as containing a fallacy. The primary use of the audit process is to ensure the validity of the logical relationships between the undesirable effects and the core problems uncovered by the ECE process.

current reality tree—A logic-based tool for using cause and effect relationships to determine root problems that cause the observed undesirable effects of the system (*APICS Dictionary*, p. 19).[10]

evaporating cloud—In theory of constraints, a logic-based tool for surfacing assumptions related to a conflict or problem. Once the assumptions are surfaced, actions to break an assumption and hence solve (evaporate) the problem can be determined (*APICS Dictionary*, p. 28).[10]

future reality tree—In the theory of constraints, a logic-based tool for constructing and testing potential solutions before implementation. The objectives are to (1) develop, expand, and complete the solution and (2) identify and solve or prevent new problems created by implementing the solution (*APICS Dictionary*, p. 35).[10]

prerequisite tree—In the theory of constraints, a logic-based tool for determining the obstacles that block implementation of a problem solution or idea. Once obstacles have been identified, objectives for overcoming obstacles can be determined (*APICS Dictionary*, p. 62).[10]

transition tree—In the theory of constraints, a logic-based tool for identifying and sequencing actions in accomplishing an objective. The transitions represent the states or stages in moving from the present situation to the desired objective (*APICS Dictionary*, p. 87).[10]

negative branch—In the theory of constraints, a logic-based tool for constructing and testing potential solutions before implementation. The objectives are to identify the impact of an action; to determine any negative consequences caused by the action; and to identify any additional actions required to achieved the desired results.

Figure 1.6   Constraints Management Thinking Processes Terms. From Cox, J., Blackstone, J.H., and Spencer, M.S., *APICS Dictionary*, 8th ed., Falls Church VA: American Production and Inventory Society, 1995. With permission.

The third element is the evaporating cloud. The conflict diagramming process was named using the metaphor of a cloud, to describe a severe problem or conflict. Traditionally, most problems are not solved satisfactorily to both sides, and their solutions represent an unpleasant compromise. The term evaporating cloud describes a situation where a solution is identified and, based on the solution, the problem simply disappears—both sides have a satisfactory or win–win solution. The cloud diagram forces the underlying assumptions surrounding the problem to surface for closer examination. Often what appears as a conflict can be researched by surfacing and challenging an assumption that can be broken, thus solving the problem. The primary use of the cloud diagramming methodology is to identify hidden assumptions which may be blocking mutually beneficial agreements (win–win solutions).

Constraints management is used as a synonym for TOC to overcome the connotations of the term theory.

WHAT IS THE NEXT STEP?

As impressive as the CM accomplishments are, caution is advisable. In order to successfully use CM, managers must understand the "big picture," the system, before they start the CM improvement process. Nothing is as damaging to a continuous improvement process as starting a project, developing enthusiasm throughout the organization, and then having to pull back because of cold feet, or a lack of understanding of the magnitude of the changes. Once burned, employees almost never regain momentum. Managers must know the CM philosophy and then decide where to use it first, second, and possibly third before starting the improvement process.

This handbook provides an understanding of the CM philosophy. In the first section a fictional company, Bob's Bolt Company, is used to demonstrate how to apply the CM concepts discussed in *The Goal* to a typical factory. Then several examples where CM concepts have been applied are examined. Finally, the next step, creating a system of continuous improvement, is discussed.

The concepts of CM have been briefly introduced in this chapter. In the remaining chapters of Section I, the development of a hands-on approach to understanding CM is continued. In Chapter 2, an overall framework of production planning and control used in modern management is built. In Chapter 3, the five-step focusing process used in CM to develop system-wide continuous improvement is discussed. In Chapter 4, the specific planning and scheduling method used in CM, drum-buffer-rope, is examined. Finally, in the last chapter of Section I, the V-A-T logical structure analysis used by managers to implement the logistics methods of CM in a specific organization is developed.

In Section II of this handbook, the concepts discussed in Section I are illustrated. Here, the application of CM to real companies is discussed using four in-depth case studies. In Chapter 6, an overview of the cases and a guide to understanding their analysis are provided. In Chapter 7, the Velmont Industries case, a V-structure, is presented. In Chapter 8, the Trane Company case, a T-structure, is presented. In Chapter 9, the Stanley Furniture Company case, an A-structure, is discussed. Also in Chapter 9, another part of the Stanley Furniture Company is used to show how two different product structures, a V-structure and an A-structure, operate together under CM.

In Section III of the handbook, the management of the factory of the future is discussed as CM techniques are combined to form a continuous improvement "learning" organization. In Chapter 10, the CM performance measurement system is discussed. Finally, in Chapter 11, the Thinking Processes to support continuous cross-functional improvement unique to CM are briefly presented.

REFERENCES

1. Skinner, W., "The Dilemma of American Operations Management: 1991," *OMA Newsletter,* Vol. 4, No. 4, Winter 1990.
2. Goldratt, E.M. and J. Cox, *The Goal: A Process of Ongoing Improvement,* revised edition. Croton-on-Hudson, NY: North River Press, 1986.
3. Goldratt, E.M., "Computerized Shop Floor Scheduling," *International Journal of Production Research,* Vol. 1, No. 2, March 1988.
4. Umble, M.M. and M. L. Srikanth, *Synchronous Manufacturing: Principles for World Class Excellence.* Cincinnati: South-Western Publishing, 1989.
5. Goldratt, E.M. and R.E. Fox, *The Race.* Croton-on-Hudson, NY: North River Press, 1986.
6. Six monographs, each containing a research paper and a case application, published by Goldratt and Fox (Vols. 1–4) and Goldratt (Vols. 5–6) from October 1987 to April 1990 as journal issues.
7. Goldratt, E.M., *What Is This Thing Called Theory of Constraints and How Should It Be Implemented?* Croton-on-Hudson, NY: North River Press, 1990.
8. Goldratt, E.M., *The Haystack Syndrome: Sifting Information Out of the Data Ocean.* Croton-on-Hudson, NY: North River Press, 1990.
9. Goldratt, E.M., "What Is the Theory of Constraints?" APICS, *The Performance Advantage,* June 1993.
10. Cox, J.F., J.H. Blackstone, and M.S. Spencer, *APICS Dictionary,* 8th ed., Falls Church VA: American Production and Inventory Society, 1995.

THE PRODUCTION PLANNING AND CONTROL FRAMEWORK

INTRODUCTION

Have you ever had to explain what is going on in a factory to a new employee, prospective customer, or your own child? To an outsider looking in, production appears to be complete chaos. Perhaps you start by explaining the various steps people are going through from order entry to customer delivery. But that seems to cause more confusion. Perhaps you explain how the product itself is put together. As a last resort, you may simply identify the different departments or functions—inspection, assembly, accounting, purchasing, and information systems. You know order exists, a reason for these people to be doing what they do, but you cannot articulate it properly. Clearly, you need a framework to build an understanding. In this chapter, a framework to help you better understand production planning and control is developed. The framework will then be used in the next chapters to master the constraints management methods.

WHAT IS THE DIFFERENCE BETWEEN A GOOD AND A SERVICE?

Of recent concern has been the reported decline in manufacturing and increase in service industries. At least from a production viewpoint, almost no difference exists between the management of a service and the management of a product. In fact, throughout this book the terms "production" and "product" will be used rather then manufacturing, as a product that is produced and delivered can be either a good or a service.

Consider the airlines, a service industry by the common understanding of the term. What happens if the airplane, crew, baggage, or fuel does not arrive at the gate at the scheduled time? Those items are clearly components that are assembled into a product. Each component has a lead time and requires a series of steps or operations to be performed. A bottleneck can occur anywhere along the series of operations and block the entire process, as anyone who has been at an airport can attest. The behavior of the production of an airline flight, a service, is the same as the behavior of the product discussed in Figure 1.2. Review the questions and answers discussed in that section. Are they also applicable to this discussion? Aren't these same questions and answers true of a restaurant or any service industry?

In fact, isn't it the case that in today's market, both physical goods and the accompanying support services are required? Think of a photocopying machine sold without a service contract or a car without a warranty. The distinction between a good and a service may be only an artificial barrier that has to be ignored to understand the production process and the impact of constraints management.

In the next section, the framework needed to better understand the role of production planning and control systems is developed. Because the early research upon which the framework was build has a focus clearly on manufacturing, all of the examples used are physical goods. The background of the authors of this handbook is also based solidly in manufacturing, and our bias, at least in examples, is in goods rather than services. However, it is important to remember that there is no real distinction between a good and a service from the production viewpoint. You can use your imagination to substitute a service application for any of the topics discussed.

THE PRODUCT-PROCESS MATRIX FRAMEWORK

In order to develop a framework, we have to look at a bit of the history of production management since most of the commonly used terms originated in these writings. Early researchers in production and operations management explored the relationship between the production planning and control systems and manufacturing strategy. Research explored the general relationship between the overall business strategy and manufacturing strategy by focusing on the physical technologies used in the production process. The early researchers

	Product structure			
	I	II	III	IV
	Low volume–low standardization, one of a kind	Multiple products, low volume	Few major products, higher volume	High volume–high standardization, commodity products
Process structure				
I Jumbled flow (job shop)	Commercial printer			
II Disconnected line flow (batch)		Heavy equipment		
III Connected line flow (assembly line)			Automobile assembly	
IV Continuous flow				Sugar refinery

Figure 2.1   Product-Process Matrix. Adapted from Hayes and Wheelwright (1979).[2]

were Hayes and Wheelwright,[1,2] Skinner,[3] Buffa,[4] and Miller.[5] Their research formed the foundation of strategic operations management thinking. The impact of this viewpoint is evident in the product-process life cycle matrix first developed by Hayes and Wheelwright in 1979. This matrix, shown in Figure 2.1, identified the relationship between the product life cycle, used primarily in marketing research, and the technological life cycle.

This relationship indicated a path from what Hayes and Wheelwright[1] called "a jumbled job shop" with low volume–low standardization products to a continuous flow with high volume–high standardization products. Two intermediate areas were identified on the matrix as disconnected line flow (batch), where multiple products with low volumes were produced, and connected line flow (assembly line), where a few major products are produced at higher volumes.

In their follow-up article, Hayes and Wheelwright[2] proposed that the product-process matrix was "...an excellent vehicle for understanding why [manufacturing] problems occur and how they can be minimized." Companies develop entry-exit strategies when managing the product offering and then must select a process strategy that will complement it. An entire industry may progress down the product-process diagonal path as the product life cycle unfolds. However, an individual company must select its technological choice at a point in time and will have considerable difficulty in repositioning itself at another point on the matrix as life cycle conditions (technology and market demand) change. Miller[5] also identified the cause of problems that plague manufacturing control systems as management's failure to fit the production system to the overall company strategy. Management faces an array of choices concerning decisions about system architecture, technology, priorities, and organization as part of the overall strategy. Miller wrote, "General management has two main roles in linking priority considerations to system design decisions. The first ensures that the manufacturing control system is structured so that priorities are set at the proper organizational level."[5] Further, "The second part of the general manager's role ensures that both the company's manufacturing control and performance measurement systems reinforce the use of priority setting mechanisms which facilitate accomplishment of competitive strategies."[5]

Various researchers linked the product-process matrix to several other functions. In 1983, Stobaugh and Telesio,[6] developed the link between the production planning and control system and the attainment of manufacturing objectives. They state that various manufacturing policies are appropriate for different product strategies and that maintaining flexibility gradually takes a backseat to achieving economies of scale (through larger, more automated plants and greater specialization of both work force and equipment). They provide several examples in which management failed to realize that a change in product strategy altered the tasks of the manufacturing system. In 1984, Hayes and Wheelwright[7] defined the concept of manufacturing strategy and placed the production planning and control system as a part of the strategy decision categories, but considered it more tactical in nature as part of the overall manufacturing infrastructure. The points on the product-process matrix were also linked to between the marketing product strategy and the manufacturing strategy also was the use of different production planning and control techniques (see Buker).[8]

Gudnason and Riis[9] refined the concept of a manufacturing strategy by identifying production technology, plant layout, production planning and control system, and organization as the key elements. They also identified two technological developments as significant to the planning and control systems. The first was the increasing use of the computer, along with a difficulty in understanding its output. The second technological development, the use of on-line terminals, with a shared database, led to the emergence of decentralization in planning and control organizations.

By the mid-1980s, researchers tested various production planning and control systems to measure their impact on the achievement of business objectives. Ritzman, King, and Krajewski,[10] also in 1984, developed a comprehensive simulation to identify which product and/or process factors, in any given manufacturing system, were critical to success. One finding was that "the repetitiveness of manufacturing operations depends to a great extent on the routings for jobs produced at feeder (fabrication) stations, the facility's emphasis on assembly rather than fabrication work (fabrication requires longer setup times and larger lot sizes), and the degree to which the routing pattern of jobs tends to resemble that of a flow shop rather than a job shop."[10] Further, in comparing material requirements planning (MRP) to kanban, they conclude, "By any relevant measure, the kanban results are exceptions. Inventory levels, for example, average just above 8 weeks of supply, as against 41 weeks for the MRP experiments..."[11] The authors conclude, however, that the manufacturing environment, not kanban, created the differences.

Fine and Hax[11] applied a modification of the product-process life cycle matrix to the Packard Electric Division of General Motors. They indicated that a successful manufacturing strategy is dependent upon the proper selection of the production planning and control system, which should be based on the plant's placement on the matrix. By the late 1980s, the role of production planning and control systems in manufacturing strategy became well accepted. Others sought to identify and to develop the relationship between production competence and performance. Using a case method, Spencer and Cox (1993)[12] sought to link the product-process matrix categories with Porter's[13] generic strategies. They classified the six companies' production processes and business strategies and then ranked the key performance areas, assessing their strengths and weaknesses. They concluded that a relationship exists between production competence and business performance.

WHICH MANUFACTURING STRATEGY IS BEST FOR EACH PRODUCT-PROCESS CATEGORY?

In 1986, Barber and Hollier[14] developed a classification of companies according to the complexity of the production control environment. They identified

six types of companies. However, their descriptions resemble repetitive manufacturers, such as tools, refrigeration equipment, pumps and valves, radiators,
and industrial batteries. Even though these companies produced repetitive
products, their research indicated that significant differences existed in the
benefits expected depending on the presence or absence of production planning and control subsystems. For example, Type 1 companies could expect to
benefit from the combination of bill of material and MRP. Type 2 companies
benefited most from capacity planning. Type 3 companies did not appear to
benefit from any subsystems other than the basic MRP calculation. Type 4
companies did not appear to benefit from any system, nor did Type 5 companies. Finally, Type 6 companies benefited most from capacity planning and
detailed scheduling subsystems.

In 1989, Kotha and Orne[15] developed a conceptual framework for the
"generic manufacturing strategies" at the strategic business unit level. They
comment on the Hayes and Wheelwright product-process life cycle matrix:
"Unfortunately, this traditional classification scheme is presently losing its
utility." They propose a multidimensional model which includes the generic
manufacturing strategies built around the process complexity, product line
complexity, and organizational scope.

As the manufacturing strategy literature developed, other authors identified
the methods used to plan and control the production process in various environments. In 1985, Fine and Hax[11] indicated that a successful manufacturing
strategy depends upon the selection of the production planning and control
system. Other researchers agreed with Fine and Hax and explored the impact
of the production planning and control system on manufacturing strategy.
Therefore, the preceding review suggests that successful manufacturing strategy depends upon the production planning and control system as well as the
selection of the manufacturing process in relation to the product strategy. It
seems that we do not yet know which manufacturing strategy is the best fit.

A NEW CONCEPTUAL FRAMEWORK

Most researchers agree on the need for a new manufacturing framework. In
Figure 2.2, we provide our conceptual framework. More complex than the
original Hayes and Wheelwright model, it includes additions based on other
researchers' work and uses as a foundation the V-A-T structure proposed by
Goldratt.[16,17] The model has eight components—customer demand characteristics, the organizational strategy, structure, and measures; logical product

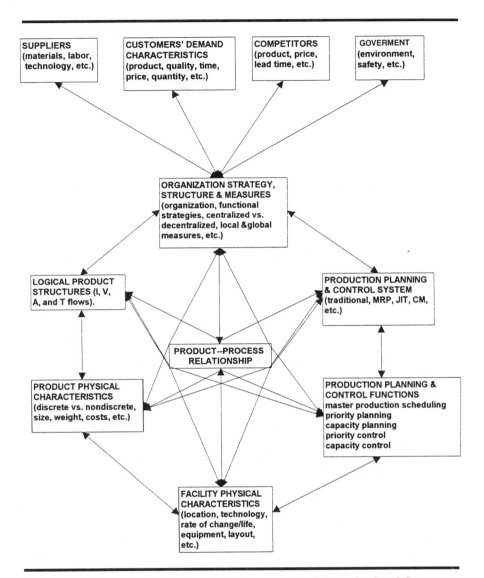

Figure 2.2   Conceptual Framework of Components of Organizational Strategy

structures; product physical characteristics; facility physical characteristics; the production planning and control system; the production planning and control functions, and the product-process relationship. While each component is discussed separately, it is important to recognize that they are components of a system and knowledge of each component for a specific environment is required to determine how to design and manage the complete system. While one characteristic of a component may dominate in a specific decision, knowledge of all of the system components and their interactions is required to effectively design and manage the complete organization. Each component of the model is discussed next.

Customers' Demand Characteristics

While several customers' demand characteristics exist and impact management decisions, only the quantity of products demanded, customer tolerance time, and product prices are discussed here. Knowing the quantities of the products demanded by the various markets is critically important in designing and managing the manufacturing and logistics processes to supply this demand. Demand knowledge is required to determine the long-term equipment, labor, and direct materials needed. Does the level of demand warrant specialized or general-purpose equipment? Is unskilled or skilled labor required? Given the product structure (I, V, A, or T), does the level of demand warrant a job shop layout, a cellular layout (for common parts), a product layout, or a continuous process layout?

Customer tolerance time is the maximum amount of time a customer is willing to wait for the fulfillment of demand. It is the time between placement of an order by a customer and the maximum point in the time the customer is willing to wait for receipt of the product. Customer tolerance time dictates how responsive a production process and logistics system has to be to support customer needs. If the production and logistics system is unable to respond in less than the customer tolerance time, then inventory must be carried in the form of raw materials, work-in-process, or finished good at the manufacturing facility or in a distribution or retail system to supply the product in less time than the customer tolerance time. A comparison of customer tolerance time and order lead time must be made to determine where inventory must be stored in a logistics chain.

The Organizational Strategy, Structure, and Measures

What are our organization and supporting function strategies? Where do we want to position our organization? Do we want to be on the cutting edge of product and/or process development or copy others' products and/or processes? Do we want to compete on price, quality, lead time, due date performance, innovativeness, field service, etc.? How do we measure our successes and failures? Do we want to compete locally, regionally, nationally, or internationally? Do we segment the market? How? How do we structure our organization to support competing on these factors, strategies, and measures? Projects? Job shop? Product? Process? We must study several other factors presented below.

Logical Product Structures

Goldratt[16,17] coined the terms I, V, A, and T to represent the major product (or product line) structures that exist in manufacturing. The term I-structure represents the simplest product structure—a dedicated fixed flow production line with no product variation. Duct tape, anvils, staples, and paper clips are examples of products where dedicated lines might be used.

The term V-structure represents a fixed flow (identical routings for all products in the family) product structure where product variation occurs. This product variation requires equipment setups and creates a divergent structure where several similar products are created from a few of the same basic raw materials. As production occurs on the raw materials at various work centers, different product characteristics are created. For example, tree trunks are cut into usable stock. This rough stock is milled then cut to dimensions (two inches by four inches, two inches by six inches, etc.), cut into lengths (e.g., 8-, 10-, and 12-feet lengths), and cured and treated. In many instances, once an operation has been performed on an item, the results are irreversible (without loss). This operation is termed a "divergent point," and divergent points are a primary characteristic of V-structures. A service example of a V-structure is the preparation of a steak—New York strip, marinated, cooked medium. Once the raw materials are selected and processed by the divergent points, it is almost impossible to reverse the process—once the New York strip steak is marinated and cooked medium, it cannot be converted to a rare New York strip steak!

The term A-structure represents a product structure where a number of raw materials, parts, subassemblies, and assemblies are processed and assembled into a few finished products. The convergence of raw materials, parts, etc. into a few assemblies and finished products is a primary characteristic of an A-structure. A second characteristic is that the operations on the parts are performed by various work centers, and a random order of operations and work centers exists for the product line. The fabrication and assembly of a pencil sharpener, a stapler, a jet engine, a riding lawn mower, etc. represent A-structures.

The term T-structure represents a product structure where either a V-structure or an A-structure has matured into a much broader product line. The increasing number of similar finished products created by combining common parts, subassemblies, and assemblies into a wide variety of products is a primary characteristic of a T-structure. The T product structure offers a number of features and options from which to choose in defining the end product. Automobiles, VCRs, pens and pencils, notebooks, and computer configurations represent product lines where similar subassemblies and assemblies or slight product variations (in color, size, etc.) are used to create a wide variety of finished products. The completion of federal tax forms, the selection of tests in a physical examination, and the selection of a meal from among the entrees, salads, desserts, vegetables, and breads in a cafeteria are service examples of a T-structure.

V-A-T analysis is discussed in detail in the next chapter. Understanding these structures and their primary characteristics is critical to the proper design and management of production processes. I-, V-, A-, and T-structures provide the foundation of the logistics paradigm in constraints management. Other logical product structures and combinations of logical product structures exist in both manufacturing and service industries.

Product Physical Characteristics

Is the product discrete or nondiscrete? Do raw materials or product characteristics dictate that the facility be located close to the raw material supply or customer demand? Does the size, the weight, the shelf life, or another characteristic dictate the facility location decision? Is the product manufactured as engineer-to-order, make-to-order, assemble-to-order, or make-to-stock?

Facility Physical Characteristics

What is the level of technology required to support the manufacture and distribution of the product to customers? What is the degree of vertical integration required to support the product? Is the capital investment enormous? Is technology cutting edge or established? Is technological change rapid or slow? Is equipment specialized, flexible, or general purpose? Have production processes stabilized? What are the environmental considerations? What are the safety considerations? Based on the product demand. is general-purpose or specialized equipment appropriate? Is a job shop or product flow layout appropriate? What is the skill base required to operate the equipment? Are specialized or multiskilled workers appropriate?

The Production Planning and Control System

Three very different production planning and control systems have emerged over several years. The first, material requirements planning, has evolved from a planning system for materials, to a materials and capacity planning system, to a closed-loop material and capacity planning and control system, to a manufacturing resource planning and control system, and more recently to an enterprise resource planning system. Numerous articles and books have been written on these planning and control systems. The second system, level scheduling and kanban, is the production planning and control portion of the Just-in-Time philosophy. While initially developed in Japan and primarily used in repetitive manufacturing environments, these systems have increased in popularity in the United States in the last decade. The third system, drum-buffer-rope, is the scheduling and control system of constraint management. It is discussed in detail in Chapter 4.

The Production Planning and Control Functions

The master scheduling function of each of the three production planning and control systems takes a somewhat different approach to planning and controlling materials and capacities. The master production scheduling function is generally at the finished product level. The remaining four functions are at the dependent demand level in each system. The structure of the materials function for each of these systems is that the bill of material for each product is exploded to determine gross material requirements and netted against existing inventories to determine manufacturing and purchasing requirements. The capacity planning function is a determination of capacity requirements to make the manufactured items and a comparison against existing labor and equipment capacities. The priority control function deals with the execution of the priority plan at each work center. The capacity control plan monitors capacity use and provides information on the need to increase or decrease work center capacity.

Product-Process Relationship

The answers to the questions in the categories discussed above cause the relationship between the product and process characteristics to emerge more clearly. The Hayes and Wheelwright framework discussed above causes the relationships between product and process characteristics to emerge more clearly. The Hayes and Wheelwright framework still provides a useful categorization of the product and process characteristics: product volume as it relates to the management methods for project, batch, repetitive, and continuous production. Now, however, we can see the reasons for this relationship to exist from the larger framework.

Of course, the answers to these questions may also cause changes to occur among categories. For example, over a period of time a company may find one product line growing more rapidly than the others. This added volume may cause the company to consider creating a separate facility devoted to the production of a single product. The logical structure may shift from an A to a T as more features and options are provided. This change may cause a shift to a new production planning and control system and a shift in other factors. The result is a shift down the product-process diagonal to a new position as production increases.

WHAT ARE THE PRODUCTION MANAGEMENT FUNCTIONS?

All of the production processes described above require appropriate management and, thus, have identifiable functions. The general functions of management from basic management texts—organizing, planning, controlling, coordinating, etc.—are too general to help understand the uniqueness of managing production versus accounting or any other area. What we need is a set of functions that are unique to production but also generalizable to all production organizations.

The origin of the five production management functions is unclear and was probably the result of managers going about their routines by trial and error. The first written description appears to be by Oliver Wight,[18] the noted production management consultant and pioneer, in his book *Production and Inventory in the Computer Age,* written in 1984. The book is organized largely around five functions common to production management. Each term[1] is specifically defined below:

- **Master Production Schedule (MPS)**—The anticipated build schedule for those items assigned to the master scheduler. The master scheduler maintains this schedule, and in turn, it becomes a set of planning numbers that drives material requirements planning. It represents what the company plans to produce expressed in specific configurations, quantities and dates. The master production schedule is not a sales forecast that represents a statement of demand. The master production schedule must take into account the forecast, the production plan, and other important considerations such as backlog, availability of material, availability of capacity, and management policies and goals (*APICS Dictionary*, p. 49).[19]

- **Priority Planning**—The function of determining what material is needed and when. Master production scheduling and material requirements planning are elements used for the planning and replanning process to maintain proper due dates on required materials (*APICS Dictionary*, p. 63).[19]

- **Capacity Planning**—The process of determining the amount of capacity required to produce in the future. This process may be performed at an aggregate or product-line level (resource planning), at the master-scheduling level (rough-cut planning), at the detailed or work-center level (capacity requirements planning) (*APICS Dictionary*, p. 11).[19]

- **Priority Control**—The process of communicating start and completion dates to manufacturing departments in order to execute a plan. The dispatch list is the tool used to provide these dates and priorities based on the current plan and status of all open orders (*APICS Dictionary*, p. 63).[19]

- **Capacity Control**—The process of measuring production output and comparing it to the capacity plan, determining if the variance exceeds preestablished limits, and taking corrective actions to get back on plan if the limits are exceeded (*APICS Dictionary*, p. 11).[19]

These five functions will be used to demonstrate how the various production management systems operate. In the next section, the development of the production planning and control systems that are used prior to constraints management is discussed.

WHAT IS A PRODUCTION MANAGEMENT SYSTEM?

Over the past 25 years, manufacturing has undergone considerable change. Before the widespread use of the computer, manufacturing was successful in providing people in the United States, Japan, and Europe an increasing supply of goods. Largely in response to the pent-up demand from World War

II and population growth, manufacturing prospered throughout the 1950s and 1960s. Manufacturing met the challenge by producing an ever-increasing quantity of products. To meet this challenge, manufacturers used a set of techniques to plan and control the production function (including reorder point, economic order quantity, min-max, two-bin system, etc.). These techniques provide large quantities of raw materials, work-in-process, and finished goals to satisfy consumer demands. However, the 1970s brought economic turbulence and crisis.

Faced with this new environment, manufacturers sought to change the basic method of managing the production function. The old techniques which had served industries so well when the focus was on increasing production became suspect when the focus shifted to decreasing costs. The first of several waves of new techniques hit manufacturers in the early 1970s. With the increasing availability of computers, several manufacturers began to explore a computer-based production planning technique called Material Requirements Planning (MRP). When successfully implemented, MRP provided companies dramatic reductions in manufacturing costs, especially in the reduction of inventory. This was especially important given the periods of inflation that occurred during the 1970s. Another key advantage of MRP was its ability to connect forecast changes to the expected levels of production required and transmit those levels throughout the organization. It was possible to identify in advance the impact of both increases and decreases in demand for a product and adjust resources accordingly. This conversion of demand into materials and capacity requirements was also a valuable feature during a period of compressed business cycles and an increasing number of new product introductions. The MRP technique was continuously refined and modified to interact with other functions, such as accounting, finance, work force planning, and budget preparation. Manufacturing seemed to be faced with only two sets of techniques from which to choose for planning and controlling the production function.

However, a third set of techniques also emerged out of the turbulence of the 1970s. After the first oil embargo, American consumers shifted their collective demand toward the consumption of smaller cars. The Japanese auto industry had supplied smaller cars to the American market for several years and was in a position to rapidly respond to this shift in demand. The American automobile companies also saw the switch but concluded that it was a temporary change. The second oil embargo ended that belief. However, the American automobile industry found it very difficult to shift to the production of small cars. Some manufacturers found it useful to establish joint ventures with Japanese firms to learn how the Japanese factories were able to make rapid modifications in production. The automobile companies and firms from the construction and farm equipment industries sought out these joint ventures, as did some machine tool makers. This investigation identified the Japanese manufacturing process called the Just-in-Time (JIT) production process as a major element of Japan's success.

Manufacturers that successfully implemented the JIT technique saw dramatic improvements in their production process. Changeovers involving large machine tools that had taken up to several days could be accomplished in a few minutes. The retooling of an entire factory could occur in less than one-third the time as under the former methods. Productivity improved and costs were reduced. Manufacturers also saw reduction in inventory levels that alone would justify the expenditure for the JIT programs. Additionally, there was an overall improvement in the quality of components and final assemblies. Many manufacturers struggled with the decision to replace MRP methods with the JIT approach. Questions arose concerning the relative merits of the production systems, the traditional precomputer techniques, MRP, and JIT. Which method was the best? When?

Manufacturers are now faced with three separate groups of techniques that have all proven to be worthwhile in planning and controlling the production function. Unfortunately, the groups of techniques are viewed as mutually exclusive to a large extent and, in fact, can cause a significant increase in cost when mixed. These techniques have been subjected to academic research to varying degrees. Each study seems to suggest the superiority of the technique being researched over the others.

CHARACTERISTICS OF CLASSICAL METHODS

It is important to explore the methods used for production planning and control prior to the development of the computer. This is to acquaint the reader with the methods used to build the classification system. It also establishes the reason why all techniques other than MRP and JIT will best fall within this classification. A description of each technique is presented, followed by a description of how the technique is used to address the five functions of production planning and control. Finally, a description of the decision rules is presented.

The period before the introduction of MRP can be described as an order launch and expedite method of production. Demand for consumer goods was primarily a make-to-stock situation, with make-to-order existing only in the industrial products sector. The retail sector ordered in relatively small quantities from the wholesale distribution system. Manufacturers replenished finished good stocks by using an order point method that had been developed in 1934 by R.H. Wilson.[20] This method established an order point which, when reached, triggered a replenishment order in the factory. The order point was calculated by taking the expected demand during the time it took to replenish the stock and adding to this a safety stock quantity to absorb any uncertainty in the lead time or demand fluctuation. The order point told the system when to make more. The quantity that was to be made was calculated using an economic order quantity formula that balanced the costs of ordering a part against the inventory holding costs. These two costs were inversely related, and where they were equal, the total cost (the costs combined) was minimized. In the factory, this method was used to replenish component parts for assembly into finished products. Sometimes a less mathematical approach was used which established production by rules of thumb. A department produced a month's worth of components one month ahead of time and placed them into a finished work-in-process area.

Although a rather simple approach, it should be noted that this method of production had several advantages. First, it met customer demand most of the time. The order point trigger released a production order in time to replenish stocks before stockouts occurred. Second, the use of an economic order quantity meant that minimum costs were planned in the system. Third, the system was quite simple to learn and manage. The triggers were largely visual signals, and the production quantities were remarkably stable over time. It should also be noted that this was the production system that was used to win both world wars and to launch during the 1950s the most explosive growth in the rate of production that had been witnessed in economic history.

In addressing the master production planning function, this method used the reorder point system to signal when an order was to be launched. Planning was rather crude in the aggregate, but the times largely required that the focus be on maximizing production. Material costs were reasonably stable, and large quantities were available at each stage of production so that expediting could be used to meet a production order's due date. The economic conditions throughout the 1950s and 1960s were a rather stable rate of growth with very little inflation. Cost management was less important than maximizing production. Master production scheduling was informal and centered around meeting customer demand.

For the capacity planning function, the use of economic order quantities presented a stable production environment. The planners were faced with an increasing level of output to be satisfied in the production plan in most cases. Economic fluctuations were absorbed through temporary layoffs. The purchasing of additional machine tools and facilities largely dictated the levels of capacity that would have had to be planned.

Priority planning was the result of the order launches being triggered by the order point calculation. Material was released into the factory to fill the released production order. There was a large amount of work-in-process inventory available to allow for expediting to occur. Most of the routings in the factory were through a job shop layout where the lead times used in the order point calculations allowed for expediting to manage the priorities.

Capacity control occurred as foremen reported their departments' production to a production control department. Overtime was scheduled either on a third shift or on weekends to produce any past due orders that may have occurred. Great pressure was exerted throughout the production system to meet the production schedules because all products could sold. An informal communication system allowed foremen to monitor feeding operations and make adjustments to their own plans as material arrived into the departments.

Priority control was established by the due dates of the production orders based on the order point calculation. Also, in the less formal method of production, the end of the month established the due date for the department. Again, the informal communication system among the production foremen helped establish the priority of production throughout the departments.

Because this was the first production system established to manage the modern industrial organization and because of its success, this method will be called the classical system. This system was able to address all five of the production functions successfully, given its operating environment. Elements of formal calculations and informal rules of thumb were used. Since the traditional system was successful in using an informal expediting approach, then it is reasonable to classify any informal approach as traditional as long as it is successful in meeting the production requirements established by management. A facility may not have an economic order quantity calculation, but it does have a rule of thumb that is used to determine the size of the product run. Some text authors have given this the formal name "period order quantity," and it is defined in *APICS Dictionary*. Likewise, any production system that establishes a due date through the use of an order point calculation will be classified as traditional. This also includes an informal system where the due date is the end of the month or week, since this also meets the period order quantity definition. Therefore, any successful production system that does not meet the classification requirements of the other three systems will be classified as using the classical method. This system is more commonly used than believed, especially in smaller companies and in companies in technically developing countries.

CHARACTERISTICS OF MRP PRODUCTION

Material requirements planning was developed in the mid-1960s. While Orlicky[21] is given credit for the first net change MRP system when he was the director of production control for the J.I. Case Company in Racine, Wisconsin, elements of MRP were developed in many different systems by many different practitioners. MRP is defined as:

> a set of techniques which uses bills of material, inventory data, and the master production schedule to calculate requirements for materials. It makes recommendations to release replenishment orders for material. Further, since it is time phased, it makes recommendations to reschedule open orders when due dates and need dates are not in phase... (*APICS Dictionary,* pp. 50–51).[19]

As the method became widely used in manufacturing, several modifications were made such that today the MRP calculation is seen simply as a scheduling tool. Other functions are addressed within the scope of the planning systems that have come to be called Manufacturing Resource Planning. Note that this new definition has the same abbreviation, MRP. For that reason, many practitioners use the term MRPII to describe this multifunctional information system. The use of MRP today implies the set of interconnected computerized programs that together address the five production planning and control functions. By reviewing the definitions of the five functions, it can be seen that the MRP techniques are intertwined within the definitions themselves.

In an MRP system, the master production scheduling function uses as input a sales forecast and actual customer orders. These inputs are converted to materials and capacity requirements and are measured against existing capacity and current inventory levels. The master schedule provides the basic inputs into the MRP system. The master schedule is usually stated in weekly time periods and can cover a horizon from 13 weeks to 78 weeks, depending on the management policy. The master schedule is usually stated in the units that the organization sells to final customers, although the bill of material can vary to support the specific needs of the organization. For example, if several options are available, the master schedule may be stated in terms of a major base component prior to the options. Almost by definition, the master schedule function is performed within the MRP system, as it generates requirements for component parts and includes the final assembly schedule for the organization.

Capacity planning within the MRP approach occurs in conjunction with the master scheduling process. Since the master schedule is forward looking as opposed to reacting to order point triggers, it translates changes in the overall economic condition into the production requirements. For example, if a recession is forecast three quarters from the current time period, then the production planning process will reflect this change as an input into the master schedule. From there, the capacity planning process will make adjustments to the levels of resources and materials. Often the two functions (master scheduling and capacity planning) are performed by the same planner at the time the master schedule is created.

Priority planning occurs within the MRP calculation itself. In order to determine when the material is required to support the master schedule, MRP passes gross requirements from assemblies to their component parts. As it does this, the time period due dates are offset by preestablished or calculated lead times as part of the internal MRP logic (lead time offset). Any existing inventory is subtracted from the gross requirement to obtain a net requirement. The net requirements for the components are then passed to the engineering routings that identify the operations and machine centers that work on the components or assemblies. In this way, time-phased requirements, consisting of start dates and due dates, are created throughout the bill of material for each part that is required and for each machine center or assembly operation that is to be used.

Priority control occurs within MRP as planned orders are launched into the manufacturing departments and material is released to match the quantities on the now opened shop orders. As actual production occurs, any deviation is identified as the progress of the order is reported to the system. As deviations occur, decisions are made to adjust the schedules or the priorities of the parts that are being made. Many MRP systems use input-output reports to identify deviations as well. These reports log the amount of work (in labor hours) entering a machine center or department against the production (in labor hours) being sent out of the department.

MRP is a set of techniques used to accomplish the five production planning and control functions. The complexity of today's products makes it virtually impossible to use the MRP method without a computer. Although the mathematical computation itself can be done manually, the other functions require the use of a computer. This is especially true in the priority planning and control functions.

CHARACTERISTICS OF JUST-IN-TIME PRODUCTION

JIT was developed in the mid-1970s by the Toyota automobile company in response to the first oil embargo. The *APICS Dictionary* defines JIT as:

> A philosophy of manufacturing based on planned elimination of all waste and continuous improvement of productivity. It encompasses the successful execution of all manufacturing activities required to produce a final product, from design engineering to delivery and including all stages of conversion from raw material onward. The primary elements of Just-in-Time are to have only the required inventory when needed; to improve quality to zero defects; to reduce lead times by reducing setup times, queue lengths, and lot sizes; to incrementally revise the operations themselves; and to accomplish these things at minimum cost. In the broad sense, it applies to all forms of manufacturing, job shops and process as well as repetitive (*APICS Dictionary*, p. 42).[19]

Researchers have identified key areas within JIT when they developed case studies concerning the Toyota system. Dr. Robert Hall (1991),[22] one of the earliest researchers of JIT, identified the following six areas:

1. Produce only what the customer desires.
2. Produce only at the rate that the customer wants the product.
3. Produce the products with perfect quality.
4. Produce instantly with no lead times.
5. Produce with no waste of labor, material, or equipment.
6. Produce with methods that demonstrate a respect for people.

Another early researcher, Dr. Richard Schonberger (1983),[23] identified 17 similar characteristics of the JIT philosophy.

Given the above descriptions of the elements of JIT, it can be seen that this is a very different approach from either the MRP method or the traditional method. However, some elements of the traditional method are integrated within the JIT system. A review of the five functions of production planning and control will highlight the JIT method as it is used in the production function.

The master scheduling function is customer driven through the use of actual orders. The time horizon is fixed for the first few months. Then changes in volume are allowed in increasing steps as a percentage of the current capacity. For example, between month three and month six, the master schedule may be increased by 20%. Between month seven and month ten, the schedule may decrease by 25%, and beyond month ten any change is accepted. Great care is used in creating a homogenized master schedule. Various different products are scheduled to match customer demand, and the mix within this schedule can change as demand changes. One day's final assembly schedule may call for the production of 200 motorcycles and 50 jet skis, for a total daily schedule of 400 labor hours. The next day's schedule may call for 150 motorcycles as demand requires, but 150 jet skis, for a total of the same 400 labor hours. The master schedule is produced manually and is distributed to the assembly areas. The first time periods are stated in terms of products to be produced per day.

The capacity planning function is performed as an integral part of the master production schedule development. The major difference between the MRP and traditional methods and JIT is the prohibition against adjusting manpower levels beyond a given level to establish a lower capacity. Toyota does use temporary workers who are subject to layoffs, but Toyota uses layoffs only as a last resort to adjust capacity. Marketing techniques such as price adjustments and the development of new markets are the methods of choice for capacity planning.

Priority planning uses the most visibly different method under the JIT approach. Material is only allowed to exist at the correct location by the design of the physical layout of the facility. Operations are closely linked, and space is simply not made available to hold idle work-in-process. A pull system is used rather than the creation of a formal schedule with due dates and start dates. As material is used on the final assembly lines to meet the master schedule requirements, a visual trigger, usually the existence of an empty container, signals the inventory replenishment. A worker simply removes a previously finished part from the preceding operation. This process continues throughout the routing until a piece is withdrawn from raw material. The removal of a piece of raw material may trigger the supplier to begin a replenishment cycle as well. Note that there is no need for any production schedule in the process.

The function of capacity control has been largely engineered into the system. The output from the final assembly schedule is measured on a daily basis at about 80% of capacity. Protective capacity is built into the schedule. Additionally, overtime is scheduled if there is a need. Since this would be considered a waste, the entire system has been designed to avoid the situation. The same is true of the fabrication areas. The priority planning approach used in JIT removes much of the risk associated with capacity control. In the Toyota system as it operates in Japan, a two-hour buffer is built in from the finish of the first shift to the start of the second shift. Employees are expected to use this buffer to finish any arrearage that might have occurred. The entire work force is cross-trained and is available to help eliminate any arrearage. Thus, capacity control is quite different than in the traditional and MRP methods.

The priority control function is also rather limited under the JIT method. Without formal schedules, there are no dates against which to measure production. Under the pull system, the lot sizes are quite small, and the lead times approach zero. Replenishment is authorized only by the physical withdrawal of a previously finished piece. The visual triggers act as the start time, and the machine process time dictates the finish time.

The five functions described above occur in the ideal JIT environment. During the transition period, larger quantities of inventory will move through the plant. None of the methods change, however. The only visible change during the transition period is the number of cards in use (called kanbans) that serve as the visual triggers from operation to operation. During the transition period, these cards are removed until the ideal state is achieved. By removing the cards from the system, constraints emerge that prohibit the removal of additional cards. The process of continual improvement is achieved by removing the constraints so that progress toward the ideal state can continue.

The five production management functions and how they are accomplished by traditional, MRP, and JIT management methods have been briefly reviewed. Before discussing how these five functions are accomplished by constraints management, it is appropriate to clear up some confusion and answer a few commonly asked questions.

HOW DOES PRODUCTION FIT INTO THE REST OF THE ORGANIZATION—OR WHERE ARE YOU ALONG THE PRODUCT-PROCESS DIAGONAL?

Ask most manufacturing managers to describe their operations and they will probably tell you that they are a job shop. The term job shop has, unfortunately, been overused. What these managers are really saying is that their production equipment is grouped by similar function. That is, all the brake presses are located together, all of the mills are located together, etc. This physical grouping, also called a functional or job shop layout, makes them a job shop according to many textbooks.

One of the unfortunate results of the use of this term is that some managers have been led to believe that JIT and constraints management (CM) methods are applicable only to repetitive manufacturing. Since some managers believe that a job shop is a different category of manufacturing than a repetitive manufacturer, they conclude that JIT and CM methods cannot be used in their factories. Kanban may not be usable in a job shop environment, but most of the elements of JIT are quite useful. Drum-buffer-rope is useful in most if not all manufacturing and service environments including projects.

We suggest that a better measure of the amount of repetitive processing that is undertaken, and thus where JIT that can be used, depends on the relative capacity assigned to the manufacture of a single product or a family of products. This relative capacity utilization will be called the "concentration ratio." The concentration ratio is the amount of available capacity divided into the amount of resources assigned to a specific family of products. If the ratio is relatively high, that is, most resources are devoted to the production of a single family of products, then the process is repetitive and should be managed accordingly. This high concentration can occur even if a factory is organized into a functional (job shop) layout.

Most American manufacturing is organized so that a relatively high amount of available capacity is assigned to a specific family of products. Therefore, even though most managers may not think so, they can manage their facilities using repetitive techniques, including JIT and CM methods. In fact, there may be a good reason why MRP shop floor control methods sometimes cause unsatisfactory results. The MRP shop floor control methods were developed for a true job shop where the concentration ratio is relatively low. As the concentration ratio increases, more and more resources are devoted to a single family of products, and MRP shop floor control techniques become less and less valuable.

Two additional results emerge from the above discussion. First, since the concentration ratio determines the amount of repetitive processing that occurs, the speed or volume of production has nothing to do with repetitiveness. Some factories produce at a rather low speed or volume, such as a farm combine factory that produces six units per week or a heavy equipment manufacturer that produces two bulldozers per day. Both may well be repetitive producers.

Second, a single factory may have both repetitive manufacturing and a job shop (nonrepetitive) operating in the same facility at the same time. It is unlikely that a single production planning and control system, be it JIT or MRP, is applicable to both environments. Therefore, the management of that kind of "joint" environment is often faced with trying to cross-breed systems, develop its own in-house system, or manage manually in order to facilitate production. What is needed is not a better MRP system but a better way to blend the appropriate components of many systems together. This is, perhaps, where CM can make its greatest impact, by creating a better way to manage a "joint" production environment.

CAN JIT OR CM BE USED IN A JOB SHOP?

If we understand the use of the concentration ratio and adopt this definition of repetitive manufacturing then the answer to this question is yes. The benefits of JIT methods are well documented. Most researchers agree that JIT requires a repetitive environment. If the concentration ratio is relatively high, there is no reason why JIT methods cannot be adopted in a functional layout (job shop). The same is true of the benefits found in applying CM principles.

In earlier days of modern production, the functional layout was considered the most efficient layout for fabrication of components. The assembly line was the most efficient layout for the assembly of components into finished products. The functional layout had its key advantage in its inherent flexibility and, therefore, its ability to keep the assembly line operating with a continuous supply of components. If one machine broke down in the drill shop, another could quickly be assigned to the production of a critical component. If one operator lacked training and was unable to manufacture a component, another operator was close by to help. The operators were able to become specialized and, thereby, their productivity increased. Supervisors could also specialize in their functions, and engineers could specialize in a particular manufacturing operation as well. Some of the specialized engineers were able to apply industrial engineering techniques, such as time and motion study, to further improve productivity. What we have is a picture of the 1950s and 1960s, when American manufacturing was in its glory.

Unfortunately, one of the problems inherent in a functional layout (job shop) is the high inventory that is required. Work-in-process components are necessary to support the assembly lines and supply operators with work to maximize efficiencies. Inventory became an increasingly expensive insurance policy in the 1970s, especially when more advanced production planning and control systems became available.

The adoption of MRP necessitated the development and use of fixed routings. The MRP calculation had to know which work center performed operations on a particular component in order to calculate the lead time. Additional advances in MRP methods were made, such as capacity calculations and shop floor scheduling methods like the critical ratio calculations. As a result, the functional layout (job shop) with its numerous work centers became overlaid with computerized bills of material, shop orders, planned orders, and routings, among other files.

The use of fixed routings reduced the flexibility of supervisors to assign operators as needed to support production disruptions. The reduction in shop flexibility, however, was greatly mitigated by MRP's ability to plan production for each work center.

Compare the functional layout (job shop) operating under the intellectual constructs required by MRP to a different type of layout where operations are organized in the same sequence as the product structure—a repetitive layout (also called a product layout). They are, in fact, the same. The only difference is that in the functional layout the physical work centers are not located together. However, except for transportation time, a functional layout is linked in the logic of the MRP calculation as if it were a repetitive layout. If managed like a repetitive layout (product layout), very little difference exists between it and the job shop using MRP, given a high concentration ratio.

This may help explain why MRP shop floor control methods are not as successful in reality as in theory. As the concentration ratio increases, more and more of the production of the same components occurs over and over, repeating itself through the master production schedule horizon. The higher the concentration ratio, the more repetitious the flow of materials, even if the functional layout remains unaltered.

IS THERE A "BEST" SYSTEM FOR EVERYONE?

Several surprises resulted from the case studies concerning how manufacturers go about managing their production processes. Probably the single most interesting finding is that companies tend to use a blend of the production planning and control techniques rather than relying on one method. Each company used one system more extensively than the others, but each also used elements from at least one other system. This was not quite as surprising given the discussion in the previous paragraphs.

For example, Motorola is known for its use of JIT methods. However, when the Motorola plant in Huntsville, Alabama was studied, it was using an MRP system for some of the production planning functions. Another example is the Trane Company, Inc., the subject of an in-depth case study in this handbook (Chapter 8). The Trane factory in Macon, Georgia was known for its use of CM, but when studied, the Trane factory was using elements of MRP and JIT in addition to CM. What has emerged is a picture of pragmatism. Superior companies use a variety of production planning and control techniques to define their own methods of managing production. What you will find in this handbook is that CM can be used by itself or along with the other systems. Constraints management is the linchpin that connects the other systems to fit the particular need of a factory.

Some rather interesting implications result for managers and researchers. Numerous researchers have discussed the advantages of one system over another. Chapter 3 in this book presents many of these discussions. A more pragmatic approach is used in reality; various elements from each of the systems are combined to develop an overall strategy for a specific manufacturing environment.

The advantages reported for each system appear legitimate in particular applications. However, the disadvantages of each system also appear to be real. The best strategy seems to be to pick the method from one of the systems whose advantages overcome another system's disadvantages, using CM as the focal point. It is a synthesis of production systems blended into an overall dynamic strategy. Since the strategy refutes the idea that one system is superior to all others under all environments, this blending strategy will be called a "constraints management synthesis." It is a blend of the classical production planning and control systems into an overall manufacturing strategy.

CREATING THE CONSTRAINTS MANAGEMENT SYNTHESIS PROCESS

The first key step in implementing the strategy is to understand the methods used in each of the three systems—MRP, CM, and JIT. The second step is to create an overall model of manufacturing as it exists in the specific factory. The third step is to extract the techniques from each of the production planning and control systems and combine them into a single production management system using CM as the focal point. Finally, the production management system is an integral part of an overall manufacturing philosophy.

In order to help understand the process of creating this synthesis, combining the five production planning and control functions (discussed in some depth in this chapter) into an overall manufacturing model will be helpful.

The master production scheduling function and the capacity planning function can be so intertwined as to become only a single ongoing function. The basic reason for this combination is the use of the production rate as the basis for both the master production schedule and capacity planning. The production rate, stated in terms of units per time period, is commonly used throughout the production planning activities. For example, if one considers the flow of production established through a factory that has a high concentration ratio, one sees a repetitive manufacturing logic.

This flow approaches that of a huge assembly line through the entire manufacturing process. The flow of material is paced in the master production schedule as a production rate. The overall capacity of the factory is determined by this production rate. If the production rate changes, the capacity changes. Only when management authorizes the production rate to change is the capacity plan changed, as communicated by the new master production schedule and now stated at a different production level per time period. The overall capacity plan is unable to change unless the production rate changes, as communicated by the new master production schedule. Capacity planning and master production scheduling become two sides of the same coin. Constraints management orchestrates the flow of material through the key resources of the facility.

The next chapter begins with an in-depth exploration of CM. The five-step focusing process and the drum-buffer-rope planning and control methods are examined.

REFERENCES

1. Hayes, R.H. and S.G. Wheelwright, "Link Manufacturing Process and Product Life Cycles," *Harvard Business Review,* 57(1), 133-140, 1979a.
2. Hayes, R.H. and S.G. Wheelwright, "The Dynamics of Process-Product Life Cycles," *Harvard Business Review,* 57(2), 127-136, 1979b.
3. Skinner, W., Manufacturing—Missing Link in Corporate Strategy," *Harvard Business Review*, 47(3), 136-141, 1969.

4. Buffa, E.S., *Meeting the Competitive Challenge — Manufacturing Strategy for U.S. Companies*, Homewood, IL: Dow Jones-Irwin Richard D. Irwin, Inc., 1984.

5. Miller, J.G., "Fit Production Systems to the Task," *Harvard Business Review*, 59(1), 145-154, 1981.

6. Stobaugh, R. and P. Telesio, "Match Manufacturing Policies and Product Strategy," *Harvard Business Review*, 61(2), 113-121, 1983.

7. Hayes, R.H. and S.G. Wheelwright, *Restoring Our Competitive Edge*, New York, NY: John Wiley & Sons, 1984.

8. Buker, D.W., "Manufacturing Strategy for Optimal Production Flow," *American Production and Inventory Control Society Synergy Conference Proceedings*, (pp. 202-205), 1984.

9. Gudnason, C.H. and J.O. Riis, "Manufacturing Strategy," *OMEGA International Journal of Management Science*, 12(6), 547-555, 1984.

10. Ritzman, L.P., B.E. King, and L.J. Krajewski, "Manufacturing Performance—Pulling the Right Levers," *Harvard Business Review*, 62(2), 143-152, 1984.

11. Fine, C.H. and A.C. Hax, Manufacturing Strategy: A Methodology and an Illustration, *Interfaces*, 15(6), 28-46, 1985.

12. Spencer, M.S. and J.F. Cox, III, "An Analysis of the Product-Process Matrix and Repetitive Manufacturing," *International Journal of Production Research*, 33(5), 1275-1294, 1995.

13. Porter, M.E., *Competitive Strategy*, New York, NY: The Free Press, 1980.

14. Barber, K.D. and R.H. Hollier, "The Effects of Computer-Aided Production-Control Systems on Defined Company Types," *International Journal of Production Research*, 24(2), 311-327, 1986.

15. Kotha, S. and D. Orne, "Generic Manufacturing Strategies: A Conceptual Synthesis," *Strategic Management Journal*, 10, 211-231, 1989.

16. Goldratt, E.M., "Gaining Momentum with Focused Productivity Improvement" Workshop. Milford, CT: Creative Output, 1987.

17. Goldratt, E.M., "Staying Ahead with a Process of Ongoing Improvement" Workshop. Milford, CT: Creative Output, 1987.

18. Wight, O.W., *Production and Inventory Management in the Computer Age*, New York, NY: Van Nostrand Reinhold, 1984.

19. Cox, J.F., J.H. Blackstone, and M.S. Spencer, *APICS Dictionary*. Falls Church, VA: American Production and Inventory Control Society, 1985.

20. Wilson, R.H., "A Scientific Routine for Stock Control," *Harvard Business Review*, 13(1), 116-128, 1934.

21. Orlicky, J., *Material Requirements Planning*, New York, NY: McGraw-Hill, 1975.

22. Hall, R.W., *Zero Inventories*, Homewood, IL: Dow Jones-Irwin, 1983.

23. Schonberger, R.J. and E.M. Knod, *Operations Management* (4th ed.). Homewood, IL: Irwin, 1991.

THE FIVE-STEP FOCUSING PROCESS

INTRODUCTION

Why hasn't this customer order been shipped yet? Why do we have such low resource utilizations and so much overtime on these same resources? Why are lead times so long? These questions are routinely asked, but largely go unanswered in most organizations. How can a firm manage its resources so that these questions disappear? The constraints management system begins with a five-step focusing process that provides the basis for effective production management. This constraints management process seeks to make sense out of the confusion that so often appears to exist in production.

The five focusing steps enables managers to plan the overall production process and focus attention on resources which create the greatest impact. This management approach has powerful implications. By understanding the five-step process, managers can challenge some of the basic principles of management.

THE TWO PREREQUISITE STEPS THAT SET THE STAGE

Constraints management has its greatest impact by directing managers to view an organization as a system as opposed to the traditional management view of optimizing the performance of each separate department. Generally, any management technique or methodology that takes a system-wide viewpoint or systems perspective is called process management. During the early days of the industrial revolution, owners/managers took a systems view in planning and controlling the organization. The early organization was relatively smaller, and all functions were located in the same geographic vicinity.

As organizations grew, a decentralized approach was taken to manage the various functions. The benefits of a decentralized approach were enormous, but there was an unforeseen cost. Owners/managers still had to exercise control over the organization; therefore, some method of feedback from the decentralized lower level managers needed to be developed. Day-to-day decisions could be delegated to lower levels, but the overall direction could not be allowed to wander or, worse, conflict among decentralized organizational units.

Two primary methods were developed in the early part of the 20th century to provide the necessary feedback and direction setting—industrial engineering and cost accounting. The benefits of applying industrial engineering measures and cost accounting measures supported decentralization. The underlying logic and assumptions of these methods are that if each area works to optimize its actions as an individual unit, then the overall performance of the organization is optimized. The sum of the decentralized actions would equal the overall performance of the entire system. This concept may sound logical, even obvious, but, as we shall see, optimizing the performance measures of each local operation does not guarantee optimum overall system-wide performance.

In Chapter 10 of this handbook, the impact of using the traditional performance measures upon an organization and the constraints management alternative are examined in detail. This chapter is devoted to mastery of the five-step focusing method for improvement. However, to understand the method, we must briefly look at what happens under the traditional methods.

THREE BRIEF EXAMPLES

The most obvious performance measure for a company is organizational profit. Profit is measured at least once a year, if not quarterly or even monthly. Profit is the result of subtracting expenses from sales. Clearly, each element of an organization should use profit as a guide to determine a course of action. The objective is to take the action that maximizes organization profits. However, the profit measure is too far removed from most of the day-to-day decisions in a production facility. While an organization measures profit monthly, quarterly, and annually for a facility, actions must be taken continually in every department. How do we determine the impact of these actions on profits? How could we allocate profit among departments/sections? It can only be done by arbitrary means. Traditionally, a surrogate—the departmental expenses measure, in most cases—has been used. If all departments minimize expenses, then won't profits be maximized? Marketing and sales can concentrate on maximizing sales. Maximized sales minus minimized costs equals maximized profits, right? Isn't this what your company does? An example might be helpful.

Example 1—It's Friday noon, and all is not well. Two more truckloads of products need to be shipped to meet the monthly shipping target for the factory. A check with the assembly department indicates that the products have already been sent to packaging. Final inspection was okay, and packaging assures us that the order will be completed by 3:00. All that remains is to ship. Shipping can easily load two trucks in 90 minutes, but the shift is over at 3:30.

It's now 4:00 and still nothing has been shipped! The products sit on the shipping dock, nicely packaged with white "release tags" flowing in the breeze. Not a living soul is around. What happened?

Inquiry on Monday reveals that the shipping department knew about the loads, had the labor, and even had the trailers available, but chose not to work the overtime that would have been required. The supervisor said that had they started to load, they would have had to finish because they can't leave a trailer half loaded and company policy won't let them ship partial loads. They would have had to work at least 45 minutes overtime. It is the end of the month and end of the quarter. It is the first time in memory that shipping will meet its expense budget—if no overtime is worked. None is, and the shipping department meets its expense budget.

A company is paid, of course, by shipping goods to customers. Shipping to customers is the obviously correct action to take. But the performance measurement system, the performance to department budget, resulted in the shipping department doing just the opposite.

Example 2—Marketing's job is to maximize sales of the most profitable products. In order to know which products are the most profitable, product cost is subtracted from product price. Since the market sets the price, we can't do much there, but if we produce those products that provide the greatest product profit margin, won't we maximize organization profits? The primary job of traditional cost accounting was to determine the cost of each product. The job was relatively easy in the early 1900s. Raw materials accounted for 50% of product cost, direct labor (based on a piece rate) accounted for 40%, and the remaining 10% was overhead (building, equipment, supervision, etc.). Approximately 90% of product cost was variable, with only 10% fixed. This overhead was allocated based on direct labor. Today, the product cost structure is quite different. Raw materials are approximately 55%, direct labor 10 to 20% (less in many industries), and the remaining 25 to 35% is overhead. Not only has overhead increased significantly, but direct labor is no longer variable. Workers are paid an hourly rate or a salary instead of being compensated based on their production rate. Direct labor still provides the allocation base for product costing, but the errors in product costing were relatively small in the 1900s, whereas today they are enormous. Think of the numerous errors that can result from incorrect product costs.

Example 3—A third example is useful in illustrating the problem of decentralization. A machine produces four components that are used in the final assembly of four products. Actual production time is two hours for each of the four components. Each component is used on one and only one end product. Because customers require all the products to be available for purchase, the assembly line produces a mix of all four products each day, hopefully in accordance with the rate of consumption by customers. The machine is a complicated one that requires a four-hour changeover between the four components. How should the parts be scheduled?

If we try to make all four products just-in-time we would have four setups or 16 hours of setup per day. Each product takes two hours of actual chip cutting or production time, for a total of eight hours. Since there are 24 hours in the day, we can just do just-in-time. If we are the inventory manager we are happy. Average inventory is one-half-day's worth or an inventory turnover of 500 times per year. A picture in the company paper, maybe even a lecture tour, could be our reward for our efforts.

But what if we are a production department supervisor. Our department performance measurement is machine utilization. The machine is considered more productive when the machine process run time is higher. Under Just-in-Time (JIT), we would have manned the machine 24 hours, but the processing time is only 8 hours. The result is a 33% machine utilization rate, which won't get our picture in the company paper. With such a low utilization, we might even lose our supervisory job!

Based on these three examples, managing the organization by our traditional methods may no longer be appropriate. Let's develop our understanding of why traditional methods may be inappropriate by answering two fundamental questions.

1. What is the real goal of the organization?
2. What is the performance measurement system that will support the attainment of the goal?

The answer to the first question is more difficult than one might think. The quick answer is to maximize profit, but that got us into trouble in the earlier examples. Some people even harbor the suspicion that the accountants can make the quarter's profit figure just about any number. Consider double-digit declining depreciation or negative net amortization, whatever they are, and the effect on the bottom line.

If the goal is not maximizing profit, then what is it? Minimizing cost? It seems that manufacturing has been on a crusade over the last 15 years to do just that. However, even though stock prices have never been higher, something seems to be very wrong. From the above examples, it seems that focusing on cost minimization is not the right answer either.

What about the answer to the second question? If traditional cost accounting or industrial engineering measures are no longer useful in operational decision making, then what? The reader will have to wait until Chapter 10 for the constraints management answer, but clearly the answers to these two prerequisite questions are critical to successful world-class management today.

WHAT IS THE IMPACT OF THE PREREQUISITE STEPS?

Thoughtful analysis is required to set the foundation for the five-step focusing process. Under constraints management, a solid foundation is required for effective execution. In the first edition of *The Goal*,[1] the goal is reported to be to "make money now and in the future." A different performance measurement system is then established to support the goal by providing local measures that identify the impact of local measures on the global goal. Local measures are those measures that managers use routinely to make day-to-day decisions.

Subsequent to the publication of the second revised edition of *The Goal*,[2] there has been additional clarification of the global goal. Some have interpreted the goal of making money now and in the future to be materialistic and self-centered, almost condoning a mercantilistic or Machiavellian approach to business. This is not a true interpretation of constraints management.

Just as in the field of economics, simplifying assumptions are also necessary here. Economics does not address all aspects of the human condition, such as making moral judgments; it only focuses on an important aspect of the problem, the production and distribution of goods and services. The same is true of the global goal under constraints management. The focus is on how business managers should operate an organization to produce goods and services. Necessary conditions must be present for the goal of an organization to be achieved. One, of course, is to obey all of the applicable laws. Another necessary condition is to operate in an ethical manner. Necessary conditions make the achievement of the goal possible. Quality products and services, equitable pay, a safe work environment, etc. are also necessary conditions. Managers should not make inappropriate assumptions about the scope or intention of constraints management simply because the necessary conditions are not expressly stated.

Given the organizational goal of making money now and in the future, and the accompanying necessary conditions, what performance measurement system can be developed to support this goal? Three measures were developed by Goldratt: throughput, inventory, and operating expense. However, caution should be exercised, as these terms have very special meanings under constraints management[2] compared to the commonly used definitions.

- **Throughput**—The *rate* at which the organization generates money through sales.
- **Inventory**—Items purchased by the organization for resale, valued at the purchased price to the firm.
- **Operating expense**—The amount of money spent by the organization to convert inventory into throughput.

For those who may be concerned at this point, the local performance measures establish a "cash" accounting system rather than a "cost" accounting system. Money flows into the organization as throughput from sales, without the accounting terminology to cloud the issue. Inventory (assets) is valued at purchase cost or the cash paid out of the organization. When it is sold, inventory is reduced by the purchased cost. Operating expense includes all other activities that use money to add value. The global goal, given the necessary conditions, is to add more money to the organization now and in the future. The rationale for these measures and their use in supporting the global goal are examined in detail in Chapter 10 of this handbook. For purposes of Section I, these definitions will suffice.

We are now in a position to examine the five focusing steps in constraints management, given the two prerequisite steps.

WHAT IS THE CONSTRAINT?

1. Identify the System Constraint

The first step is to identify the constraint in the system that limits throughput. Let's look at Figure 3.1. Recall from Chapter 1 the power of the single resource that limits the output of the entire system. In this case, it is operation 20. Clearly, management must focus on the constraint, as the performance

<u>EXAMPLE</u>

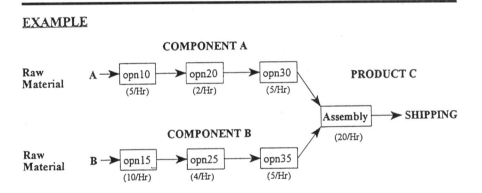

( ) = Quantity that can be produced per hour

<u>QUIZ</u>

- What is the maximum output of product C? Why?

- What happens if operation (opn) 25 increases its output to 10/hr through increased efficiency? Why?

- What happens to the output if more than 2/hour are released into operations 10 and 15? Why?

- What happens if operation 10 breaks down for 4 hours? Why?

- Is the result different from a breakdown of operation 30? Why?

Figure 3.1   A Simple Production Example and Quiz

of the entire system depends on it. Many types of constraints exist. Let's look at the most obvious, a physical resource, first.

Any system, whether a biological or production system, must have something that limits its growth. Otherwise the system would grow unbounded and consume all in its path. The need for a limit to a biological system is more obvious than in a production system. If a production system had no constraint, the organization would make an unlimited profit. Unrestrained growth leads to monopolistic power and to a chat with the U.S. Department of Justice. There must be a constraint in any system. If there is no physical constraint, an organization could produce more than it can sell, which would make the market for the product the constraint.

If there is a constraint somewhere, then how can we identify it? Sometimes it appears that there are many bottlenecks or the bottleneck appears to move from resource to resource. In most cases, this is caused by scheduling and lot size policies. For any product line there is almost always a single constraint, and it usually stabilizes when the production system is managed properly using constraints management.

Identifying the bottleneck takes some work. The V-A-T logical structure analysis, which provides a systematic structured approach, will be discussed in Chapter 5. In the meantime, however, the best approach is often to go to the production floor and ask knowledgeable employees (schedulers, expediters, production supervisors) about the flow of products through the production process. These employees can almost always point to one or two resources as the constraint. A look at the overtime records will also be illuminating.

HOW DO WE USE THE CONSTRAINT TO ACHIEVE THE ORGANIZATION GOAL OF MAKING MORE MONEY?

2. Decide How to Exploit the System Constraint

The second step is to decide how to exploit the constraint. The word "exploit" is an interesting one. The implications are to take advantage of something. Here it means taking advantage of the existing capacity at the constraint, which is often wasted by making and selling the wrong products and by improper policies and procedures for scheduling and controlling the constraint. Since any time lost at the constraint is production lost to the whole system, it is critical to make the right mix of products and to keep the flow to the constraint at all times.

There are some more obvious ways to maintain the production flow through the constraint. Remember, by definition the constraint limits the performance of the production system. It therefore does not have the existing capacity to process all products. The first suggestion is to ensure the constraint is working on the right products to maximize profits. After determining the mix, it is a good idea to randomly observe the constraint and see what the employee(s) and resource are doing. Is the resource working continuously without interruption? Often such common policy decisions as when employees take lunch, breaks, clean up after work, and change shifts can be modified to add more processing time to the constraint each day.

The Japanese have developed excellent tools for improving production. Use the JIT tools focusing primarily on the constraint. Reduce setup times on the constraint to zero, apply quality control procedures (where appropriate) to eliminate defects entering and after the constraint, and adopt preventive maintenance practices on the constraint. All of these exploit the capacity of the constraint to increase profit.

WHAT DO NONCONSTRAINT RESOURCES DO?

Step 3. Subordinate All Else to the Constraint of the System

The third step is to subordinate all other activities to the constraint. This is quite difficult to implement. It is at this step that some managers lose their courage. What exactly does this step mean? It means what it says—all other activities are subordinate to the constraint. It is the most difficult step because it flies in the face of most traditional managerial policies and practices and local measures.

Three common performance measurement systems block implementation of this step. However, in order to understand the five-step focusing process, let's briefly look at how these three measurement systems block implementation.

The first performance measurement system contains the productivity measures used in industrial engineering. Look at Figure 3.1 again, but now assume that we are looking at separate departments where the different operations are located. Operations 10 and 15 are performed on rough mills located in the milling department. Operation 20 is a grinding operation performed in the grinding department, which is located in a separate building. Operation 25 is a drilling operation, etc. Clearly, operation 20 is the constraint.

This may be clear to us, but it is not nearly as clear to the supervisors and/ or managers of the different departments. If the performance of the production facility is measured using productivity measures such as equipment utilization or percent of labor efficiency (standard direct labor hours divided by direct labor hours), what will happen? The constraint is unaffected since, using step 2, it is producing more than before. But what happens at operation 10? If material is released only at the rate of two units per hour, productivity will be 40% (two units divided by five units per hour capacity). No medal for that supervisor. In fact, all of the operations except the constraint now have idle capacity by design. Managers might try to reassign workers to balance the workload, by using computer reports common under material requirements planning, but the closer we get to a perfect balance, the closer we get to creating a new constraint somewhere else. A problem will clearly be created if the management and measurement of nonconstraint departments are not subordinated to the physical constraint. Work-in-process inventories and lead times increase. Product mix priorities are diluted.

Cost accounting is the second performance measurement system that can block this step. What happens when the cost system picks up the hours being generated by the constraint department versus all other departments. All nonconstraint departments have idle capacity; otherwise, they themselves would be the constraint. Traditional cost accounting assigns overhead to the products with the most direct labor hours. What will happen to the perceived costs of those productd? Pressure increases to reduce the labor content of these products; therefore, by focusing improvement efforts on reducing product cost, additional idle capacity is created at nonconstraints.

The third performance measurement system is the human relations system itself. With pay commonly tied to individual output (individual incentive pay systems), what happens to the pay of the operator at the constraint compared to the pay for all nonconstraint (occasionally idle) operators? Will the nonconstraint operators be satisfied to subordinate their pay to the performance of the constraint operators? How will the constraint workers feel when they come under the scrutiny of managers, even overlapping lunch periods and minimizing setups, when other workers in the facility have built-in spare time? What happens if nonconstraint workers exceed the constraint pace? Their pay increases, but so does the work-in-process inventory, without any increase in throughput. If the plant is unionized, then how can the contract be rewritten to support the operation of an around-the-clock constraint and not punish nonconstraint workers? Step 3 is rather difficult to implement, but it must be implemented in order for constraints management to succeed.

HOW DO WE MAKE MORE MONEY IN THE FUTURE?

Step 4. Elevate the Constraint of the System

The fourth step is to elevate the constraint. Elevate in this case means to increase the capacity to a higher level. This step can be confused with step 2, to exploit. We evaluate and can now determine if we want to add capacity to the constraint, thereby increasing its capacity. This is commonly done by making modifications to existing equipment and using higher speeds and feeds or by adding equipment to the department. Some production routed to the constraint could be off-loaded to alternative equipment as well.

WHAT NEXT?

Step 5. If in Step 4 the Constraint Is Broken, Go to Step 1

The fifth and last step of the five-step focusing process is to prevent inertia from stopping the process of continuous improvement. This step allows the achievement of a balance not present under JIT methods. Managers balance the cost of making a change against the profits coming from the change. Rather than changing the production system on an hoc, unplanned basis, changes under constraints management are focused on the department or operation that provides true continuous improvement for the organization. Continuous improvement is defined as movement toward the organizational goal. The organizational goal is to make more money now and in the future. Managers are cautioned not to allow the profits achieved under constraints management to create managerial inertia. Rather, they are challenged to plan continuously improving the performance of the entire organization by focusing on actions to improve the performance of the constraint.

If the constraint is broken in step 4, it is important not to let managerial inertia set in. Instead, go back to step 1 and start the process over. The five-step focusing process allows management to think, plan, and then do and check. It is a more systematic process of continuous improvement.

Implementing step 4 may cause another problem, however. If the capacity at the constraint is elevated beyond a certain level, then that constraint may have been broken, but a new constraint may appear somewhere else in the system. This action will cause a massive reallocation of the constraints management effort.

In addition to redesigning the scheduling system used under constraints management, consider the changes made under step 3. The human relations performance changes, engineering design changes, make versus buy decisions change, product pricing and mix change, commission structures change, and quality improvement efforts must be structured to support the new constraint. Most policies and procedures must be reexamined. Clearly, this step cannot be undertaken lightly. However, instead of running into the problem of the unplanned impact of a continuous improvement program such as JIT, we can now prepare ahead of time and even elect not to shift the constraint until a more appropriate time. Determining where the constraint should be for an organization is a strategic decision. What does a process of continuous improvement really mean for an organization?

ANSWERS TO FREQUENTLY ASKED QUESTIONS

A few questions are frequently asked concerning the five-step focusing process. The answers to these questions should improve your overall understanding of constraints management.

1. What if There Is More than One Constraint?

In reality, it is extremely rare to have more than one constraint in a given product flow. In most cases where two interactive resource constraints exist, a policy (e.g., lot sizing, sequencing of setups, etc.) creates the dependency between the resources. In a few cases, true interactive constraints exist in a product flow. The easiest solution to interactive constraints is to determine where logically the constraint should be in the production system and elevate/eliminate the other constraint. It is quite difficult to manage interactive constraints effectively unless sophisticated scheduling software is utilized.

2. What Do I Do if I Have a Floating Constraint?

A floating constraint is defined as a situation where the constraint moves from one department to another department periodically. A floating constraint appears to occur at a rather high frequency. However, upon closer investigation, actual floating constraints are also quite rare. Most floating constraints are caused by management policies such as using large lot sizes, running to high machine utilizations, sequencing work orders at different production departments to minimize setup times, assuming transfer batch is equal to process batch (explained below), variable manning of equipment, etc. In Chapter 5, the use of a production framework called V-A-T analysis will be discussed. V-A-T analysis allows managers to classify production flows so that the five-step focusing process and the scheduling methods (discussed in the next chapter) can be used. In the V-A-T framework, the logical product structure combines the components' routings and the bill of material. Based on V-A-T analysis and the logical product structure, when the drum-buffer-rope scheduling method is used a single department almost always emerges as the constraint.

What managers are seeing as a floating constraint is really the result of large amounts (in lot sizes) of parts attempting to flow through a production system, which causes timing problems as the whole lot moves from operation to operation. It is like watching a snake eat. The lump moves slowly down the snake. At any point in time, a different point along the snake that appears to be the constraint, but it is really the lot size itself that is causing the floating constraint. The floating constraint is the result of a timing problem and not a time problem.

Shop schedule expediters and supervisors can tell you which department causes them the most trouble over time. This department is most likely the actual constraint. The constraint is probably a production department that is used to produce most of the components routed through the shop. The parts contend for the constraint resource. Contention is made worse by the use of large lot sizes which, just like the snake, try to force a large lot through the department all at once. The drum-buffer-rope scheduling methodology uses two lot sizes—the process batch and the transfer batch.

The process batch is the amount of parts run on a machine at one time. If long setups exist on the constraint, the constraint process batch should be large. If little to no setups occur on the constraint, then the constraint process batch should be small. The processing batch at nonconstraints should not be small enough to create a new constraint but should be large enough to idle other resources so that they do not become interactive constraints. The transfer batch is the amount of materials that moves between departments. The transfer batch should not be equal to the process batch. Overlapping of operations eliminates interactive constraints (caused by timing). Small transfer batches create smooth product flows through the facility and reduce lead times significantly.

3. The Constraint Has to Be a Production Department, Doesn't It?

Actually, most constraints are not physical constraints at all! A constraint is anything that limits the throughput of a system. The way constraints management works is most easily seen by looking at a physical resource, as is the case in this chapter, but the constraint may not be a physical operation at all. As we saw in step 2 of the five-step focusing process, managerial policies can easily become the actual constraint that limits output.

Various functions can create different types of constraints. For example, transportation policies or actual shortages of equipment can limit material flow. Conveyor systems often become constraints. They limit the amount of material flowing into and out of the production system and between stations. Occasionally they are designed to stop when a workstation is down. The conveyor actually accumulates the downtime or setup time of all equipment it feeds, thus severely limiting the throughput of the entire system. These are logistical constraints. Another constraint can be a limit on obtaining the raw material needed in a production process, a supply constraint. Even the market itself can become a constraint if consumers are unable or unwilling to buy at the current price all of the goods or services the organization has the capacity to produce. Any function or policy can block improvements. If the constraint is a physical constraint, then the five-step focusing process can be effectively used to manage the system so that throughput is maximized in support of attaining the global goal. If another function or a policy is the constraint, then the current reality tree (discussed in Chapter 11) of the thinking process may be required to surface the problem.

4. What Is the Most Common Type of Constraint?

This is a difficult question. Many constraints management researchers believe that managerial policies, rather than a physical constraint, are more likely to be the actual organizational constraint. Managers are frequently limited by what they think are good policies and procedures which actually support only local goals at the cost of failing to support the global goal. Overtime is a common example. How many organizations limit the amount of overtime that can be worked, especially at the end of a quarter? In some cases, overtime requires the approval of a top manager. Yet, by limiting overtime at a true physical constraint, the throughput of the entire system is less, plus inventory of non-constraint components builds up and unnecessary operating expenses are added. A valued customer may be lost by not using overtime at nonconstraint departments to meet a delivery commitment.

Step 2 says to exploit the constraint. Overtime could be a relatively small addition to operating expense that would produce relatively high throughput. Upon closer examination, managers find that a great deal of their policies and procedures actually act to limit throughput, because most policies and procedures are written to support traditional performance measurement systems.

Some researchers suggest the market itself is often the constraint, especially in difficult economic times. The market, acting through supply and demand, establishes the equilibrium between market price and quantity. Capacity of physical plant and equipment actually operates as a step function. Once a drill is in place, for example, it is available 24 hours a day, 7 days a week. The quantity of labor seems to be the asset that actually defines the capacity of most organizations. By adding one more unit of labor, a relatively small addition to operating expense, an organization can produce an amount of throughput for sale. If the market can be persuaded to buy the additional units, then the throughput (sales quantity times market price minus variable costs) flows directly to the organization's bottom line. The ability to balance market price and throughput capacity may be the activity that actually limits the system. Traditional cost accounting practices for basing product price on product costs limit our ability to effectively eliminate market constraints.

5. What Is the Difference Between Exploiting and Elevating the Constraint?

Constraints management is not an exact science or a recipe. The idea is to use constraints management as a process to continuously improve a business by more effective management. A good way to understand the difference between exploiting and elevating is to look at the effect on the capacity of the system. By "exploiting" the constraint, short-term actions are taken that add capacity to an existing resource. An example would be to better schedule the flow of material to and through the resource or to ensure that the resource is running during lunch and break times.

"Elevating" the constraint lifts the overall resource capacity to a higher level, usually by adding capital equipment. The equipment need not be new. An alternative routing across nonconstraint equipment may be used, freeing up constraint capacity, or a fixture might be purchased for the constraint to reduce the setup time for some parts.

6. How Do I Implement "Subordination"?

How badly do you want to implement constraints management? As discussed previously, this step is probably the most difficult to implement because it flies against conventional wisdom (e.g., common practice). However, by taking the time to develop an understanding of constraints management, it becomes clear that it is common sense. Many companies, such as the ones described in Section II of this handbook, elect to run two parallel performance systems, one based on common practice and the other based on common sense. It is unlikely that cost accounting can be quickly replaced by constraints management accounting methods. The financial community uses the traditional accounting approaches, as do various governmental agencies. There is, however, a major difference between financial accounting and cost and managerial accounting. The government and the financial community are very restrictive in terms of what a company can do in financial accounting. Cost and managerial accounting are less controlled by external forces. With computer-based accounting systems, it is really quite easy to maintain parallel systems. Constraints management accounting can be used to make managerial decisions and traditional financial accounting can be used for historical and governmental purposes.

The same is true of industrial engineering performance measures. There is no governmental requirement to continue to use traditional productivity measures. However, there is some advantage to maintaining continuity for historical comparisons. Additionally, a huge organization cannot implement constraints management accounting all at once. Over time, various organizational units can implement constraints management decision making. However, as soon as the traditional measures can be abandoned, they should. Conflicting measures can create chaos. Top managers must keep in mind that the power of constraints management is in the impact on the organization's bottom line and that all other measures are subordinate.

7. Would We Ever Not Want to Go to Step 5?

Step 5 is a transitional step that can loop back to step 1. In the next two chapters, the drum-buffer-rope scheduling method and the V-A-T framework will be presented. Considerable work goes into establishing the flow of parts through a production system operating under constraints management. Additionally, marketing and sales focus on exploiting the constraint by identifying desirable products and markets based on the constraint. Engineering determines alternate routing to off-load the constraint. Quality ensures that constraint parts are not later converted to defects. Care must be taken not to disrupt that flow in an unplanned or poorly planned manner. If the decision is made to elevate the capacity at the constraint beyond a certain point, another resource may become the new constraint. The capacity may also be elevated to such a level that the constraint is broken and the market itself or another function (engineering, transportation, etc.) becomes the new constraint. In any case, managers should plan their actions carefully beforehand.

If the constraint is broken, a new set of control points can emerge and, at a minimum, the focus of the scheduling process will have to shift to the new constraint. If the constraint is broken in step 4, then preparations must be made ahead of time to maintain the flow of throughput. Breaking the constraint and going back to step 1, of course, is the essence of the continuous improvement process. It is important to have an understanding of the situation and a plan of action before step 4 is taken.

Step 5 cautions not to let inertia or the fear or cost of change stop us from the continuous improvement process. Under constraints management, it's *ready, aim, fire*—in that order.

AN EXAMPLE OF THE FIVE-STEP FOCUSING PROCESS

Bob's Bolt Company (fictitious) will be used to illustrate the application of the five-step focusing process to productivity improvement efforts. While the focusing process starts in the factory, administrative or service functions interface with and affect the performance of a factory. First, the five-step focusing process will be traced and illustrated; then, present performance improvements that resulted from the focusing process will be presented.

Process management is an excellent technique for documenting and defining administrative as well as manufacturing processes. The objective of process management is to improve productivity by clearly defining process outcomes and customers and coordinating the process activities that cross organizational boundaries.

The process management approach has had greater success in improving productivity in manufacturing than in service tasks because service tasks often lack clearly defined activities, boundaries, and ownership and detailed process diagrams. Without clearly defined processes, service managers may find it difficult to establish control points and to define and implement improvement measures (discussed in Chapters 4 and 5). Consequently, managers may take actions to improve the productivity of an isolated activity or department but harm the overall process and organizational productivity.

The approach illustrated here shows how general process management can be enhanced with the five-step focusing process. The purpose of the five-step focusing process is to enable a system to undergo continuous improvement by identifying and managing the system's constraints.

To reiterate, the five-step focusing process consists of the following steps:

1. **Identify the system constraint(s)**—In this step, the manager, using process management analysis, studies the process flow to find the resource that limits the throughput of the entire system. In simple structure, a system constraint may be identified by a long queue of work or long processing times.
2. **Decide how to exploit the system constraint(s)**—In this step, decisions must be made about how to modify or redesign the tasks of the constraint so that work can be performed more effectively and efficiently.
3. **Subordinate all else to the constraint(s) of the system**—In this step, management directs all of its efforts toward improving the performance of the constraining resource and any other resources that directly affect the constraining resource.
4. **Elevate the constraint(s) of the system**—In this step, additional capacity is obtained that will increase (elevate) the overall throughput of the constraint.
5. **If in step 4 a constraint is broken, go to step 1, but do not allow inertia to cause a system constraint**—In this step, the process of ongoing improvement is implemented. As a result of the focusing process, the improvement of the original constraint may cause a different resource to become the constraint. Inertia may blind management from taking steps to improve the system's throughput now limited by the new constraint.

Company Background

About 95% of the products produced at Bob's Bolt's factory are make-to-stock. Yet, total customer lead time was quoted at 45 workdays, which included 25 days for manufacturing lead times used in the material requirements planning calculation and 20 days for administration and distribution. The administration function includes order acceptance, order entry, and credit approval for each customer order. The distribution function includes the traffic and shipping department. This department consolidates customer orders and makes shipping arrangements. About 82% of all orders receive credit approval and are shipped within 30 days of receipt.

Bob also reports that the factory is often subject to the "end-of-the-month" syndrome, when overtime is used to fulfill shipping targets during the last week of the accounting period and workers are "laid off" during the first week of the new accounting period because no inventory is left in the production system to complete shipments. The "end-of-the-month" effect causes an increase in costs as overtime is used at month's end, and workers on lay-off are unproductive at the beginning of the month until called back to work after the first workdays in the new month.

The Original Order Entry Function

Prior to applying constraints management and the five-step focusing process, the order entry process consisted of the following steps: (1) orders are received from sales representatives by fax, mail, or telephone into a centralized order entry department; (2) once an order is received, it is entered into a sales order system; (3) a credit approval process is undertaken, and shipping is authorized only when approval is obtained; (4) the master production scheduling function is then notified of the new order by its access to the sales order system, and an allocation from the warehouse stock is made; (5) the master production schedule is modified by replacing the forecast quantity with the new actual order quantity; (6) the final assembly schedule is modified as required; (7) the traffic department consolidates customer orders, arranges for trucking, and schedules delivery to a customer's location.

A team of trained personnel applied the five-step focusing process to identify the limiting task in the order entry process. They viewed order entry as a batch manufacturing process using a first-come, first-served scheduling rule. Queuing in front of each step increases the lead times, as is the case with manufacturing operations. Each order waits its turn in line at each department.

Within the order entry system, the five-step focusing process was used to identify the new system constraint. By examining the queues at each step in the process, the credit approval operation was identified as the constraining activity. These queues resulted from varying approval procedures that require a review of each customer's credit history. In addition, these procedures vary by the size of the credit requested. For example, an order for $5000 requires different approval steps than an order for over $500,000.

In step 2, the constraint is exploited. Actions were implemented within the credit approval function to streamline the process based on the overall dollar volume of an order. For example, the credit approval process for orders from existing customers with a good credit rating was streamlined by identifying these orders early in the process and giving them a high processing priority. As a result of this process, the amount of order backlog at each workstation was reduced, thereby reducing queues in the order entry process and total administrative lead time.

In the third step, all other activities are subordinated to the constraining task and supporting activities. Management attention was directed toward the credit approval function and inventory allocation through the master production schedule. In this case, the order entry process interfaces with the inventory allocation and the master production schedule by issuing permission to consolidate inventory items for delivery. The management of this organizational boundary is important to the overall reduction of customer lead time. The goals of the other departments are given less emphasis.

In step 4, the constraint is elevated; that is, the constraint is eliminated through management actions, either by a change in existing procedures or the acquisition of new technologies. By implementing new procedures that better integrated order entry and inventory management, it was also determined that the allocation of warehouse inventory could be made prior to final credit approval without disrupting the master schedule. Step 5 determined that the constraint in the order entry process had been broken again. As a result, the five-step focusing process led the improvement team to identify the traffic-shipping function as the new constraint.

The New Traffic-Shipping Function

The five-step focusing process was again used to improve the overall productivity of Bob's administrative functions. It was determined that the new constraint was now located in the company's traffic and warehouse departments. When credit approval was granted for a customer order, the approval triggered a warehouse inventory allocation subject to master scheduling review. Then, the order was sent to the traffic department. This centralized department was responsible for deciding which orders should be consolidated based on total weight and geographic destination. Then, the traffic department sent the consolidated orders to the warehouse supervisor, who was responsible for picking the orders from finished goods inventory, packaging the orders, and preparing them for shipping.

In step 2, the constraint is exploited. In this case, the physical movement of products within the warehouse was reduced by prestaging shipments using outlines of trailers drawn on the warehouse floor. This new procedure reduced the chance of damage to the products during handling and reduced the time needed to load a truck. The warehouse shipping procedures were also streamlined by using a chalkboard to communicate the status of the daily shipping schedule on an ongoing basis. This schedule provided the necessary information for taking immediate actions as conditions changed.

As was the case previously, all other administrative activities were subordinated to the traffic and warehouse departments in step 3. In the fourth step, improvements in the traffic department were made by changing its overall consolidation strategy. First, the department reduced the number of motor carriers used by the factory and negotiated to have the trucking companies perform the consolidation at their facilities. The factory also implemented a program that allowed shipping as little as a 200-cubic-foot order from the warehouse. Now, the trucking companies are responsible for consolidating a broad range of available shipments and creating full truck loads for specific geographic areas. These changes reduced lead times from customer order acceptance to order receipt significantly.

In step 5, it was determined that the constraint was again broken and the process was started over. This time, the new constraint was identified as the assembly lines. Management determined that it could manage the internal operations of the factory using drum-buffer-rope (discussed in the next chapter) and decided to keep the assembly lines as the constraint. The five-step process would again be used, but now it is focused on the manufacturing activities rather than the administrative functions. Applying the five-step focusing process to the administrative activities provided significant improvements in lead time, which increased product demand.

The five-step focusing process provides a way to operationalize the process management approach to increase organization productivity. As a result of the actions taken, over 60% of the orders at the factory are now shipped within seven days of receipt. Previously, the goal of the factory was to ship orders in 12 to 15 days, and 20% of all orders were late. Additionally, 95% of the orders are now shipped within 15 days of receipt. Under emergency conditions, an order received on one day can be shipped by 10:00 the next morning. (Ultimately all orders should be shipped in this time frame.) The factory work force now has a more evenly distributed workload over the accounting period. Administrative costs have been reduced accordingly. Finally, Bob's Bolts has used improved customer service as a competitive advantage.

Notice how the five-step focusing process forces the management team to view the entire system in this case the production and order entry functions but focuses efforts on the constraint in order to make system improvements. As a result, improvements in the overall process can be obtained even if nonconstraining activities have to subordinate their activities to support the constraint. The five-step process has applicability in manufacturing and service activities. Improvements in service activities, especially those that support manufacturing activities, can have as significant an impact in overall performance as changes in the manufacturing process itself.

WHERE TO START THE PROCESS

Process management is a productivity improvement technique that can be applied to both manufacturing and administrative tasks. The objective is to define and coordinate the interdependent tasks that may cross departments. By understanding the work flow within an organization, improved productivity can be gained. This chapter focused on achieving productivity improvement through process management techniques more effectively by implementing the five-step focusing process. The key contribution of the five-step process is the identification of the resources that, when managed correctly, can yield the highest productivity gains for the organization. Improvement efforts can then be focused on these resources.

The next chapter examines the scheduling methods used in constraints management to implement the five-step focusing process. In the last chapter of Section I, the five-step focusing process is linked with the scheduling methodology to provide an overall framework that is used to manage a production system using constraints management.

REFERENCES

1. Goldratt, E.M. and J. Cox, *The Goal: Excellence in Manufacturing.* Croton-on-Hudson, NY: North River Press, 1984.
2. Goldratt, E.M. and J. Cox, *The Goal: A Process of Ongoing Improvement,* 2nd edition. Croton-on-Hudson, NY: North River Press, 1994.

THE DRUM-BUFFER-ROPE SCHEDULING METHOD

INTRODUCTION

The previous chapter provided a way of looking at the production system as a whole rather than as a series of separate operations. In this chapter, the drum-buffer-rope scheduling method of constraints management (CM) is examined. In the next chapter, the V-A-T analysis method of locating the key operations or control points will be discussed. These control points provide managers the ability to obtain the greatest reward for their scheduling efforts. Together, these three chapters form the basic foundation for implementing the logistics concepts of CM.

First, a detailed step-by-step process for developing a production schedule is provided. Buffer management, the method of executing the production schedule, is then discussed. The use of the drum-buffer-rope scheduling method and buffer management allows managers to implement the concepts discussed in the book *The Goal*[1] and to achieve remarkable results similar to those presented in the first chapter of this book.

THE DRUM—THE MASTER PRODUCTION SCHEDULE FOR THE CONSTRAINT—IS THE KEY

Probably because CM is evolving, the confusion that emerges concerning the development of the master production schedule (MPS) if one reads only *The Goal*[1] was inevitable. The book was not written as a textbook to implement a scheduling procedure, but rather to present the underlying management constructs. Even subsequent articles discussing the drum-buffer-rope method have been somewhat unclear as to the relationship between the MPS and scheduling of the constraint. This chapter is based on both an examination of the articles discussing CM, The Haystack Syndrome and the investigation of companies that have implemented the drum-buffer-rope procedure.

Constraints management requires directing management's attention to the constraining resource. Actually, the constraint may be a resource or a management policy or procedure, as discussed in the previous chapter. In this chapter, we will assume that the constraint is a physical resource. This does not mean that the customer requirements are suddenly being ignored. It does, however, mean developing and managing the MPS differently in fulfilling customer requirements.

Under material requirements planning (MRP), the MPS is a statement of time-phased production requirements for items to meet customer requirements. These MPS items are usually the end items or products. The end item bill of material specifies the components and raw materials. The MPS can schedule major assemblies which are then scheduled for assembly to meet customer requirements, as in an assemble-to-order environment. In many traditional cases, the bill of material structures dictate that the MPS be developed at somewhat lower levels, especially where there is a dominant assembly line(s) present, as in the automotive, furniture, consumer products, and other repetitive industries.

Under CM, the MPS is the plan of action at the constraint. This is not as radical as it may seem when one considers that almost no one master schedules the true end items in the bill of material. Management usually aims the MPS at a lower level in the item bill of material. The MPS is scheduled to the bill-of-material level which best fits the product structure.

Additionally, this does not mean that under CM management is no longer interested in the shipment or delivery of its products. The attainment of the global goals requires shipping end items just as before. However, under CM, the shipping of end items is orchestrated to maximize net profits and return on investment, and maintain cash flow, by maximizing throughput from the constraint. As will be seen in the following section, the MPS is a plan to maximize throughput from the constraint. The shipment of end items is accomplished through the final assembly schedule and shipping schedule, which are derived from the drum (the constraint schedule).

A Constraints Management MPS Example

Let's again use Bob's Bolt Company as an example and assume a job shop environment consisting of eight work centers. In celebration of the bicentennial of the United States, Bob has the shop produce three patriotic end items (a line of red bolts, white bolts, and blue bolts) from four types of purchased raw material (W, X, Y, and Z), on one 40-hour-per-week shift. Demand each week for the next two weeks is 40 reds, 53 whites, and 40 blues. Figure 4.1 shows the shop layout (a), the bills of material (b), and the routings (c). The layout shows one press, two grinders, two polishers, two drills, one assembly, and a shipping area. The bill of material for red shows that raw materials X and Y are assembled to form AR, which is processed to provide a red bolt. White and blue bolts are constructed according to their respective bills of material.

Inventory on hand is as follows: 25 pieces of assembly ARs in finished work-in-process (WIP), 10 pieces of component W at operation 40, and 15 pieces of component Z at operation 30. In order to more clearly see the relationships, the bills of material and routings have been combined into a single diagram, often called the logical product structure (see Figure 4.2).

Under a traditional performance measurement system and using MRP, the following MPS could be developed: ship 40 reds, 53 whites, and 40 blues on Friday of week 1, followed by 40 reds, 53 whites, and 40 blues on Friday of week 2. MRP calculations do the rest to create a production plan for all components, with start and due dates for all work centers. A capacity requirements plan analysis is presented in Figure 4.3. This analysis indicates that there is enough capacity at all machine centers to execute the MPS, although the press is a concern.

Using the five-step focusing process from the previous chapter, the constraint is identified as the press. In this case, when the time for needed setups is included, the press exceeds 100% of available capacity. The next step is to exploit the constraint so that throughput is maximized. This requires the development of the MPS to strategically plan component production at the press. A Gantt chart of the constraint illustrates the principles involved in MPS development.

We now know that the construction of the MPS is based on the press. The routings indicate that the only components needing the press are component BR for operation 10, component Z for operation 30, and the raw material for the finished blue product for its operation 40. The MPS will consist of those components that are routed across the press, the constraint, rather than the end red, white, and blue products. The location of existing WIP inventory must also be examined to see how soon the press can begin to process material. The press must operate 100% of the time to maximize its and the system's throughput.

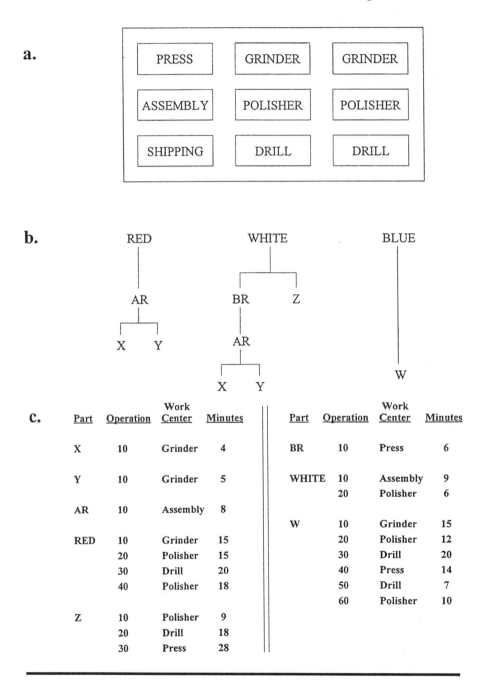

Figure 4.1   (a) Layout, (b) Bills of Material, and (c) Routings. (Data based on the simulation problem of Goldratt, E.M., *Production: The TOC Way.* New Haven, CT: Avraham Y. Goldratt Institute, 1996)

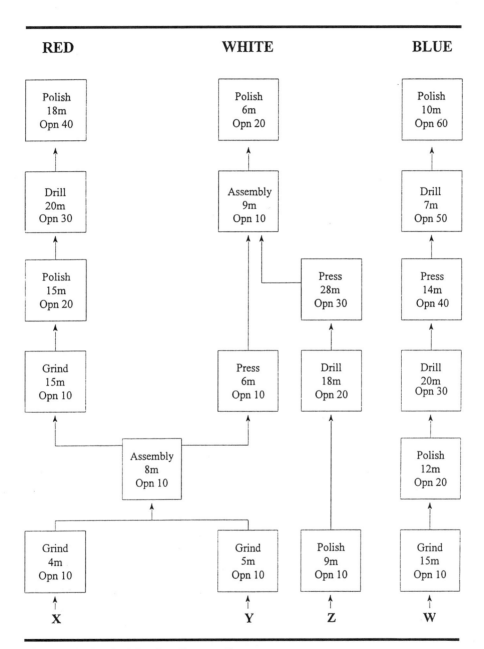

Figure 4.2 Logical Product-Process Structure

| Part/Rqmt | Process Minutes on Work Center | | | | |
	Press	Grinders	Polishers	Drills	Assembly
Red	0	24	33	20	8
40 pieces	0	960	1320	800	320
White	34	9	15	18	17
53 pieces	1802	477	795	954	901
Blue	14	15	22	27	0
40 pieces	560	600	880	1080	0
Total Process Minutes	2362	2037	2995	2834	1221
Machines[1] Available	1	2	2	2	1
Utilization (%)	98	42	62	59	51

[1] 2400 minutes are available for one machine and 4800 minutes are available for two machines.

Figure 4.3   Capacity Analysis by Work Center

Priority Decisions

The WIP inventory allows us to decide what to run first among the three components at the press work center. We still maintain a link to the end items under CM even though the MPS no longer consists of the end items themselves. Since 10 Fs were promised for midweek shipment based on press availability and 15 Ds were promised for later delivery, the choice is easy. The choice could be dictated by the amount of processing time required for each end item's total number of components. For example, the white end item needs both BRs and Zs to ship. Each white end item needs 6 minutes on the press to make one BR and 28 minutes to make one Z, for a total of 34 minutes.

On the other hand, the blue end items need only 14 minutes of processing time on the press. Each white end item requires more than twice the processing time at the constraint. Depending on the constraint (profit) margins involved, it may well be to our advantage to produce the blue items first so that if anything does go wrong in the shop, the end items with the lower constraint margin will be delayed. This calculation requires a twist, however. Under CM, constraint margins (see Chapter 3) are calculated as the selling price minus raw material costs. Neither direct labor nor overhead is subtracted from price, as both are assumed to be fixed organizational costs in the short run.

Let's assume the profit contribution from the white items is $145 and the profit contributions for the blue items is only $115. It would seem that we should produce the white bolts first because they provide a greater profit contribution. Since we want to exploit the constraint, we must consider the constraint's processing time in the equation. The contribution per constraint minute for the white bolt line is $4.26 ($145 divided by 34 minutes required at the constraint). The contribution per constraint minute for the blue bolt line is $8.21 ($115 divided by 14 minutes at the constraint). We obtain almost twice the profit with blue products for each minute of processing time at the constraint! Thus, we would schedule the components going to the blue end item ahead of the white end item's components as long as the constraint is not waiting for material.

The Drum—The Master Production Schedule of the Constraint

The purpose of the drum under CM is to facilitate the attainment of management's objective of maximizing throughput from the constraint. In our example, the constraint must operate 100% of the time (step 2 in the five-step process). The MPS is developed across time at the beginning of day 1. Based on the above calculation, blue items should be made first. The inventory position allows us to begin the MPS with a 15-minute setup at the beginning of day 1 and produce ten Ws (see Figure 4.4).

After the ten Ws are produced, the press can be set up to run the components going to the white end item (bolts). Of the two components, BRs take 6 minutes and Zs take 28 minutes. Since it is early in the week, it doesn't matter which of the components with the shortest processing time first. Should something go wrong, later in the week we would be in a better position to produce components that take a longer time last. For example, if we were to lose 60 minutes toward the end of the week we would lose only two end items (28 minutes for each component) rather than ten end items (6 minutes for each component). Having established the importance of producing all 40 of the blue end items, we schedule the remaining 30 Ws. Next, we schedule the remaining BRs. Finally, we schedule the press to run the last Zs to finish

Part	Operation	Run Qty	Minutes Per Unit	Run Time	Set-up Time	Total Minutes	Total Hours	Cumulative Hours
W	30	10	14	140	15	155	2:35	2:35
BR	10	15	6	90	15	105	1:45	4:20
Z	30	15	28	420	15	435	7:15	11:35
W	30	30	14	420	15	435	7:15	18:50
BR	10	38	6	228	15	243	4:03	22:53
Z	30	38	28	1064	15	1079	17:59	40:52

GANTT CHART -- PRESS

Figure 4.4   Drum or Master Production Schedule Computations for the Press

the shipping requirements for the week. Remember, the constraint is the drum in drum-buffer-rope scheduling. The schedule for the press is the rate at which the drum consumes work. Note that there is still time remaining in the week for production at the constraint. We would like to replenish the WIP inventory we had at the beginning of the week so that we will be able to fully use the constraint at the beginning of week 2. We finish the first week with 40 Ws in the MPS to support the next week's shipping requirements.

An end item that does not have any components which are routed across the constraint can be thought of as a "free good." The red end item is a "free good." Since the red item does not require any of the constraint resources, ample capacity exists to produce and ship the 40 pieces demanded by this Friday. The red product may be scheduled evenly (daily) across the week, with enough time allowed at the end of the week to give shipping a small time buffer for safety. This, then, is the "free good" MPS under CM (see Figure 4.5).

A Summary of the Constraints Management Approach to the Master Production Schedule

The development of the MPS consists of the following:

1. Determine the constraint (using capacity analysis) (Figure 4.3).
2. Determine which components are routed across the constraint.
3. Use the contribution per constraint minute for each product to determine the constraint's priorities.
4. Use these priorities to build a Gantt chart of the constraint's use.
5. Schedule any end items that do not contain components routed across the constraint (free goods) evenly in the MPS.
6. Develop a material release schedule by scheduling backward from the constraint time. The time offset should provide enough time for the materials to get from raw material release point through the nonconstraint work centers.
7. Develop the shipping schedule by scheduling forward from the constraint time plus the shipping buffer. The time offset (amount of time added to the constraint time) or shipping buffer should provide ample time for the queue, setup and processing, and move times for all nonconstraint operations between the constraint and the shipping dock.

HOW TO DEVELOP THE MATERIAL RELEASE SCHEDULE

How do we know that the right quantity of material will be available to support the constraint's schedule? From the constraint schedule developed

Event	Parts	Actions	Minutes	Press Start Times	Press Finish Times	Products
1	W	Set-up	15	Day 1 7:00	Day 1 7:15	Blue
2	W	Run 10	140	Day 1 7:15	Day 1 9:35	
3	BR	Set-up	15	Day 1 9:35	Day 1 9:50	White
4	BR	Run 15	90	Day 1 9:50	Day 1 11:20	
5	Z	Set-up	15	Day 1 11:20	Day 1 11:35	White
6	Z	Run 15	420	Day 1 11:35	Day 2 10:35	
7	W	Set-up	15	Day 2 10:35	Day 2 10:50	Blue
8	W	Run 30	420	Day 2 10:50	Day 3 9:50	
9	BR	Set-up	15	Day 3 9:50	Day 3 10:05	White
10	BR	Run 38	228	Day 3 10:05	Day 3 2:53	
11	Z	Set-up	15	Day 3 2:53	Day 3 3:08	White
12	Z	Run 38	1064	Day 3 3:08	Day 6 7:52	
13	W	Set-up	15	Day 6 7:52	Day 6 8:07	Blue
14	W	Run 40	560	Day 6 8:22	Day 7 9:42	

Figure 4.5   Drum or Master Production Schedule for the Press

above, the quantity and release time of the raw material can be planned. This is the rope component of drum-buffer-rope scheduling and is similar to the netting logic in MRP. Material is released to the gating operations (operation 10) at the same rate (standard hours) as production occurs at the constraint (standard hours are consumed). If more material than is consumed by the constraint is released, it would simply build up as unneeded WIP inventory. Any lesser amount released than is consumed by the constraint would starve the constraint. The material release schedule should provide a time buffer between material release and the constraint. This is accomplished by using back scheduling from the Gantt chart of the constraint to the first operations where material is released. By subtracting a time buffer from the constraint schedule (MPS), material release at the first operation is determined. The time buffer should be much larger than the sum of the processing times at all operations between release and the constraint, should be large enough to allow most operations to work on other components first, and should be large enough to absorb the unplanned disruptions that are common in a production environment. A rule of thumb in CM is to start with a buffer that is five times the sum of the setup and processing times of operations between material release and the constraint and make adjustments to the buffer as actual production occurs. As in the case of MRP, the release date for raw materials is determined by the lead times in the routings. The time buffer is an additional amount of lead-time offset allowed in the development of the material release schedule (see Figures 4.6 and 4.7).

The third step of subordinating all the operations to the constraint is now implemented. A degree of discomfort is likely to occur because many of the traditional performance measurements are abandoned at this step. All other operations are subordinated to the pace of the constraint by linking material release schedules to the constraint MPS. Unlike MRP, there is no need to calculate the start and due dates for any other operation. As raw material is released in accordance with the material release schedule, production occurs at the nonconstraint operations. If material is available at a work center, then it is worked on. If material is not available, two conditions can exist. First, if the constraint MPS indicates that material should be in the constraint buffer and it is missing, then expediting at nonconstraints occurs. Second, if the constraint has ample material in its buffer, then there is no immediate need for the nonconstraint operations to produce. The nonconstraints should wait for work to come to them. This is part of buffer management.

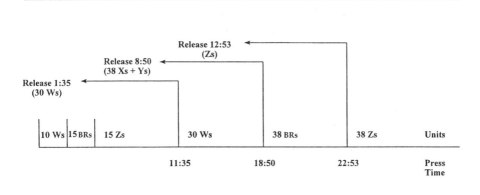

Figure 4.6a   Graphic Material Release Schedule (Backward Scheduled) for Drum (Press) (a ten-hour constraint buffer is subtracted from the drum schedule)

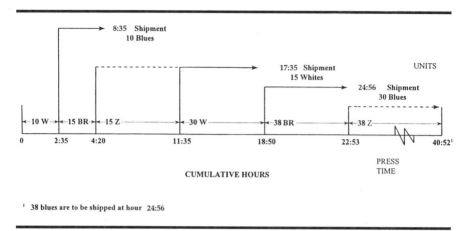

¹ **38 blues are to be shipped at hour 24:56**

Figure 4.6b   Graphic Shipping Schedule (Forward Scheduled) for Drum (Press) (a 6-hour shipping buffer is added to the constraint schedule)

HOW TO DEVELOP THE SHIPPING SCHEDULE

The shipping buffer (see Figure 4.6b) allows its safety time to absorb any instability in the actual operating environment. As was the case in determining a material release schedule using the constraint schedule to construct a shipping schedule, a time buffer is added to the calculated constraint schedule (based on the constraint Gantt chart) to determine the shipment date. This forward time offset is called the shipping buffer.

Many products are constructed by assembling components or subassemblies. An assembly schedule is used to orchestrate the convergence of components and subassemblies into the finished product. An assembly schedule can also be derived based on the constraint schedule by backward scheduling the assembly operation of the nonconstraint and constraint parts from the constraint part time to raw material release for the nonconstraint part by an assembly time buffer. Thus far we have discussed how to determine the constraint schedule, the raw material release schedule, the shipping schedule, and the assembly schedule. These schedules are derived from the capacities and priorities established for the constraint.

Keep in mind that the constraint MPS contains only components routed across the constraint. However, the plant MPS must schedule all end items manufactured in the facility. In addition to pulling components from the constraints to the shipping area, the shipping schedule is used to schedule the nonconstraint products. Like other production planning systems, there is

Event	MPS	Action	Release Time	
a. Constraint				
1	Run 10 Ws	Released	—	
2	Run 15 BRs	Released	—	
3	Run 15 Zs	Released	—	
4	Run 30 Ws	Release 30 Ws	Day 1	7:50
5	Run 38 BRs	Release 38 Xs	Day 2	7:50
		Release 38 Ys	Day 2	7:50
6	Run 38 Zs	Release 38 Zs	Day 2	11:53
7	Run 40 Ws	Release 40 Ws	Day 4	2:00
b. Free Goods				
1	Run 10 Reds	Released	—	
2	Run 10 Reds	Release 10 Xs	Day 1	11:00
		Release 10 Ys	Day 1	11:00
3	Run 10 Reds	Release 10 Xs	Day 2	11:00
		Release 10 Ys	Day 2	11:00
4	Run 10 Reds	Release 10 Ys	Day 3	11:00

Work starts at 7:00 am each day, lunch is 12:00 to 1:00 and work ends at 4:00 pm each day. Calculations are based on a 10 hours constraint buffer and a 12 hours shipping buffer. Process batch is not equal to transfer batch. Transfer batches are small to allow for overlapping processes.

Figure 4.7   Constraint (a) and Free Goods (b) Material Release Schedule Example

only one plant MPS under CM. This plant MPS is based on the constraint schedule. The constraint schedule then drives subordinate schedules for material release, assembly, and shipping to orchestrate production. A "free good" (or nonconstraint product) shipping schedule is then developed so that these goods will not create a new constraint due to poor scheduling (see Figures 4.6b and 4.8).

A 6-hour shipping buffer is added to the constraint finish time for the last constraint part in a given product (white) to determine the shipping time. For free goods, the shipping time can be determined based on customer need date. The shipping buffer (12-hours for free good, in this example) is then subtracted from the shipping time to determine the raw material release time. A second approach is to determine when raw materials can be released without creating a new constraint. The shipping buffer is then added to the release time to determine shipping time.

BUFFER MANAGEMENT—THE EXECUTION PHASE

Whatever is planned has to be executed. Successful execution is performed by managing the various buffers. Luckily, buffer management, whether the constraint buffer, shipping buffer, or assembly buffer, is the same. Another example from Bob's Bolt Company will be used to illustrate the management of the constraint buffer (see Figure 4.8).

Let's assume for this example that the constraint is machine center 3 (MC3). An MPS for one week was developed using drum-buffer-rope scheduling, as shown in Figure 4.8a. Note that the production hours required to produce the quantities has been determined from the MPS. Also, assume an eight-hour workday and that setups for components A and B take one hour each and for components C and D two hours each. Monday's plan reads: first, setup (one hour) and produce component A (two hours); second, setup (one hour) and run component B for the rest of the day (four hours).

In order to see how buffer management works, visualize the MPS as a Gantt chart for MC3 (Figure 4.8b). Compare the MPS with this Gantt chart. They are identical; the Gantt chart is simply a visual representation of the MPS.

Now assume that we have decided upon a three-day constraint buffer. We can visualize the constraint buffer as depicted in Figure 4.8c. Notice that the three days are now described as three regions along the horizontal axis. Time is depicted along the vertical axis beginning with the present time zero at the origin and increasing through the production day to hour eight. Starting on day one at time zero, the Gantt chart is converted to this graph of the buffer.

This WIP inventory is what we would expect to see if we went out and actually looked at the physical components sitting in front of MC3. We would see the amounts of components A, B, and C sitting by MC3. We omit depicting the setup times since we can't see any movement of material during a setup. All of the setup times are seen only as the components sitting in front of the machine center awaiting their processing. Many companies that use CM actually paint lines on the floor to represent the buffer and its

Release Parts	Constraint Finish Times		Products	Shipping Times	
a. Constraint Products					
10 Ws	Day 1	9:35	10 Blues	Day 2	7:35
15 BRs	Day 1	11:20			
15 Zs	Day 2	10:35	15 Whites	Day 3	8:35
30 Ws	Day 3	9:50	30 Blues	Day 4	7:50
38 BRs	Day 3	2:53			
38 Zs	Day 6	7:52	38 Whites	Day 6	2:52
b. Free Goods					
Release Parts	Raw Material Release Times		Products	Shipping Times	
10 Xs 10 Ys	Released		10 Reds	Day 2	700
10 Xs 10 Ys	Day 1	11:00	10 Reds	Day 7	7:00
10 Xs 10 Ys	Day 2	11:00	10 Reds	Day 4	7:00
10 Xs 10 Ys	Day 3	11:00	10 Reds	Day 5	7:00

Figure 4.8 Shipping Schedule for Constraint Products (a) (based on a 6-hour buffer added to the constraint finish time) and for "Free Goods" (b) (based on a 12-hour buffer added to raw material release).

regions as a way to to reinforce the importance of getting work orders positioned in front of the constraint machine center.

The buffer is established in this manner. As time passes and setups and production take place, the composition of the buffer changes. Raw materials are released based on consumption at the constraint. However, a three-day buffer is still expected buffer as new components arrive from feeding nonconstraint work centers. In our example, the changeover to component A is completed on Monday morning and production begins. At the end of Monday, the composition of the three-day buffer has changed. We are in the production run of component B. Region 1 now contains all Bs, and region 2 should contain the last of the Bs in the run and most of the Cs. In fact, region 2 has simply now become region 1, while region 3 is now region 2, and region 3 has been built containing Thursday's run (from Figure 4.9b) of the remaining Cs and part of the Ds. All components in region 1 and most components in region 2 should be present at MC3.

The Day-to-Day Operation of the Buffer

Suppose we examine the buffer on Tuesday morning at 7:00 and find no components in the constraint buffer. The plan called for Bs and Cs and at least some Ds, but there are none. What will happen? The constraint (MC3) will be unable to produce anything until some Bs show up from the feeding operation. We not only lose the downtime at the bottleneck, but the system loses the profits that could have been made had the buffer contained components. Clearly, when the buffer is empty, it is too late to take action. A good supervisor protects the future, at least a few days, to ensure that the constraint never makes an unplanned stop based on missing orders. The buffer contains the future work scheduled on the constraint.

The buffer management gives a supervisor an early warning system that extraordinary actions must be taken to prevent constraint downtime. Sometime Monday, there should have been more component Bs coming from the feeding operations. A good supervisor would compare the constraint schedule to the components in the constraint buffer and notice the missing components. He or she would look for them or, at least, make a few telephone calls to obtain a promise time. In this manner, if the components were not going to arrive as planned, then alternatives could be generated to keep the constraint active. Perhaps the feeding operation making Bs should work overtime or perhaps component C can be expedited early.

What would happen if we used a ten-day buffer instead of a three-day buffer just to be safe? Nothing would happen unless there was enough instability in our production system to require that size of a buffer. We would know if holes (missing parts) started to appear in the buffer and expediting had to occur. Suppose no holes appear in the buffer. Over a period of time, we would conclude that a ten-day buffer is simply too big. We would then reduce the buffer size by one or two days and wait to see if any holes showed up in the region of the reduced buffer. If no holes appeared, then we would again reduce the buffer size. Ultimately, holes will appear, given the variability inherent in any production system.

a. Schedule of Orders for MC 3 (Constraint)

DAY	ORDER	PROCESSING HOURS	DAY	ORDER	PROCESSING HOURS
MON	A	2	THU	C	3
	B	4		D	3
TUE	B	8	FRI	D	2
				E	4
WED	B	1			
	C	5			

b. Production Sequence for MC3

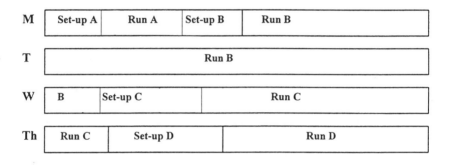

c. 3-Day Buffer For MC3

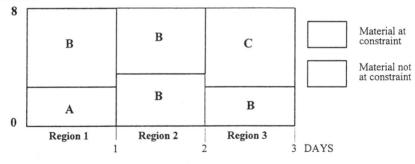

Figure 4.9   Buffer Management

The appropriate size of any buffer is established by setting a fairly large buffer first and then reducing it until holes appear in region 1. The three-region approach can be used to help in fine-tuning. Instead of visualizing the regions as one day each, the buffer size is simply broken into three regions. For example, through trial and error we set the constraint buffer at six days. Each region now consists of two days of components. A useful rule of thumb is to set the buffer so that, on average, one-half of the total buffer is full of components awaiting their turn at the constraint and one-half is empty. The empty portion is a visual signal to alert the supervisor to take action if no replacement components arrive as time passes. Phone calls to feeding operations can usually elicit a promise or alert the supervisor to take extraordinary actions.

Once the various buffers are established, the continuous improvement process of buffer management is initiated. The planned reductions in the buffer size are used to focus improvement efforts. The reduction in the buffer size triggers additional expediting as holes appear more frequently in the buffer. Investigation into the causes of these holes allows management to focus improvement activities. This is very similar to the continuous improvement approach some Just-in-Time (JIT) companies use in a planned removal of kanbans.

CM uses visual triggers—the buffers—similar to the kanban triggers used in JIT. All of the buffers work in the same way as described above. These buffers are time buffers. The buffers can be created manually by Gantt charts, as described previously, or by the lead time offset in MRP, which is discussed in the next section.

CONSTRAINTS MANAGEMENT IN AN MRP SYSTEM

In order to see the relationship between the MRP system and drum-buffer-rope scheduling, a distinction must be made concerning the operating environment. Both systems, as well as JIT, operate differently in a job shop versus a repetitive environment. As discussed in Chapter 2, in a traditional job shop, machine centers are grouped by similar function. That is, drills are grouped in one department, mills in another, and lathes in a third. MRP was developed to plan and control production in this environment. Lead times in a job shop are calculated to reflect processing times across different operations, move times, inspection times, and queuing. The queue time is frequently quoted at being around 85 to 90% of total lead time. Queue is simply the wait time involved as a part awaits its turn for the operation to be performed. It is this queuing time, however, that provides the flexibility to manage a part's relative priority, which is so critical to MRP's scheduling success.

According to the *APICS Dictionary* (1995, p. 63),[4] priority planning is "the function of determining what material is needed when. Master production scheduling and material requirements planning are the elements used for the planning and replanning process in order to maintain proper due dates on required materials". Priority control is "the process of communicating start and completion dates to manufacturing departments in order to execute a plan" (p. 63). Thus, without a method to change a part's priority, no plan can be executed under an MRP system. The method of changing priority is through the management of the queue in a job shop. As MRP changes a due date, the floor supervisor can accommodate the change by expediting or de-expediting as required. Since the processing time is fixed, the supervisor can manage the change by expanding or contracting the queue relative to other orders.

In a repetitive environment, equipment is arranged in the order in which the operations are to be performed. That is, if a part is routed so that a milling operation is required first, followed by a grinding operation and a drilling operation, then the machine centers are physically arranged in that manner: the mill, followed by the grinder, followed by the drill. The result is that move times are greatly reduced, as are queue times, as the material flows smoothly through the plant. This physical arrangement has several advantages over the traditional job shop arrangement under many conditions. However, one major drawback has been the virtual elimination of queues; thus, the floor supervisor is unable to change order priorities to meet MRP schedule changes. In a repetitive environment, the physical flow of material across the various machine centers is often accomplished by conveyor; therefore, the interruption of the flow of a part is very difficult and often very costly. Once a lot is started through a repetitive process, it often must simply be completed before other material is launched. The JIT production techniques closely follow this type of repetitive process environment. This layout and its management are very different from the traditional job shop arrangement. Thus, any production planning system must recognize the differences between environments and respond accordingly. It may be this failure to recognize the differences between the job shop and repetitive environments that has caused some of the disappointments attributed to a MRP system. It may be that the MRP system is performing its function, but management fails to recognize that it is planning in the wrong environment.

Using the Drum-Buffer-Rope Technique
for a Job Shop with MRP

Earlier we discussed the crucial role a constraint plays in production. The constraint resource governs both the flow of material through the plant and the resulting inventory level. Thus the statement "an hour lost at the constraint is lost for the total system" makes sense. This recognition leads to understanding that an hour lost on a nonconstraint is minor and that the utilization of a nonconstraint depends on the system's constraint. Thus, if one believes that the insights presented in *The Goal* are true, then the resulting management concepts make sense as well. The question then becomes how a manager can make use of these insights in an MRP system.

The first step is to identify the constraint resource in the shop. Most MRP systems have a work force/machine load report as part of the normal computer output. This report takes the quantities to be produced on all machine centers and multiples them by the standard hours for labor, or by the actual cycle time standards for the machines. The report is normally in weekly time periods and covers at least 12 weeks into the future. By knowing the work force shift policy (one shift is equivalent to 40 hours per worker per week or two shifts to 80 hours per worker, etc.), the capacity can be roughly determined. Very few systems recognize setup times as part of this calculation; therefore, the result will probably understate the actual loads. In the absence of this kind of report, the master scheduler probably has a set of machines that are usually reviewed for changes in capacity. These resources are most likely the constraint resource(s), and talking to the shop supervisors and expeditors can quickly help pinpoint likely candidates. Since the use of CM is a continuous improvement philosophy, it is okay to make an error at this stage, because the real constraint will surface during the drum-buffer-rope process.

Once identified, the actual cycle time of the resource or the actual production of the worker determines the throughput for the entire factory. This is the drum. The master scheduler must have courage to implement this pace as the controlling element in the creation of the MPS. A prerequisite to this implementation is an understanding of the CM concepts and an overall agreement by management that this approach is correct. Assuming that this has been done, the master scheduler now must accept no additional requirements beyond what can be produced at the constraint during a time period. The master scheduler must be given access to the shop production schedule for the constraint as well as the production reports. A master schedule based on the constraint must be constructed first, and the master schedule that feeds the MRP system is then created from this constraint schedule.

The task of the master scheduler is to develop a finite schedule for the constraint. The original MRP shop orders are used as the menu from which the finite schedule is prepared. A visual Gantt chart or a spread sheet can be helpful. Care must be taken in handling the current and past due orders, as these orders will show a sizable overload situation in most systems. Some overtime on the constraint may be required simply to catch up. Remember that 85 to 90% of the calculated lead times used by the MRP system to create the original shop orders was queue time; therefore, a considerable amount of excess time exists. Using the actual production rate of the constraint resource is the best way to develop the overall constraint load (capacity) and the schedule that results from fitting the shop orders to that load. An additional advantage to developing this schedule with a team is that certain economies can be obtained. For example, sequencing similar orders across the constraint can eliminate unneeded setups. Once the finite schedule has been created, the next step is the creation of the master schedule. End items which contain constraint components must appear in the master schedule in a manner that is consistent with the quantities and the timing of the Gantt constraint schedule. The primary function of the master schedule is to ensure that the material flow supports the constraint production. The constraint schedule must be converted to the shipping schedule. The free goods must also be scheduled using the shipping schedule. This shipping schedule is the MRP master schedule. The execution of the schedule is a separate job. The two are not equal.

At this point, the next step, creating buffers, may dispel any apprehension the master scheduler may feel. Buffers are viewed differently under CM than under MRP. Buffering often is made more difficult if management has embraced JIT. Under CM, inventory is viewed as a strategic weapon that is needed to manage the statistical fluctuations that will occur in any dependent event system. In *The Goal*,[1] these statistical fluctuations are highlighted by the analogy of the dice and matchstick game played by the scout troop. The only way to avoid statistical fluctuations is to eliminate variability in all elements of the production process. Under JIT, considerable effort is devoted to this task. If this variability can be eliminated, then JIT will outperform most other systems, but until this perfect world can be achieved, the advantages of using strategically positioned buffers cannot be overstated.

The first buffer is placed in the shipping area. This is the time offset from the constraint schedule to the promised ship date. For free goods, it is the time offset from shipdate to raw-material release. In most MRP systems, some time buffer already exists since no differentiation is made within a week's time bucket. That is, promised dates on Friday appear to MRP's weekly bucket as due on Monday. Most manufacturing systems compensate for this weekly period by supplementing MRP reports with daily shipping schedules. Expediting from the final assembly area is geared toward meeting this ship schedule. Often the final assembly schedule is created to support this ship schedule. The size of the shipping buffer depends on the variability that must be managed.

Again, because CM is a continuous improvement process, buffer sizes do not require a scientific calculation. In fact, over time, this buffer size will decrease as improvements are made. Unless the MRP system is operating in less than weekly buckets, no action is necessary to create this buffer. Coordination between the shipping scheduler and the final assembly scheduler can quickly create this buffer if it does not already exist. The major difference between MRP and drum-buffer-rope schedules is that the shipping schedule is determined based on the constraint schedule.

The next buffer to be created is an assembly buffer. It protects the shipping schedule against external and internal disruptions. Assembly buffers for purchase components are time buffers created to ensure that delivery of components does not interrupt the assembly schedule. Again, there will be statistical fluctuation within the routing of nonconstraint parts that may cause downtime at assembly. Until the shop can eliminate this variability, the line must be protected by this assembly buffer. In reality, most assembly systems already use an offset to compensate for this variability. The second use of the assembly buffer is to protect the assembly line from variability within the factory. This variability occurs throughout the plant, but the line is protected from any part routed across the constraint by the constraint buffer discussed below. The flow must be protected from shortages of any parts not routed across any constraint. Again, unless an MRP system has time buckets of less than a week, this buffer already exists, but in an ill-managed manner. A part's lead time may place it in an earlier time bucket as it nets from higher levels that do not recognize a Friday from a Monday within the same week. The shop production schedule which executes the plan from MRP does appear to include this time, as most lead time is stated in days. Keep in mind that the execution of the plan is very different from the creation of the plan. It is an easy task to add one or two days of administration lead time to all parts not routed across the constraint. Most MRP lead time calculation subroutines have this parameter. However, care must be exercised to add this time to only those parts that enter the final assembly area from the manufacturing area rather than to all parts throughout the plant. A review of the routing file to identify those parts is required.

The buffer to be implemented is at the constraint resource. As before, this is a time buffer. Typically, lead times are carried within MRP at the part level as a parameter in the part master file. Access to this file occurs during the lead time calculation as part of the MRP explosion routine. Again, it is a rather easy task to identify those parts routed across the constraint from the routing file. Once the constraint has been identified, a time buffer is created by adding administrative days to the routing.

It may appear that the creation of these buffers is an exercise in adding waste. However, it it is important to remember that these buffers are strategically placed to protect against statistical fluctuations. Excess lead time already exists within the planning side of MRP. All that is necessary is to recognize the buffers' value in the execution phase. The overall WIP inventory level will be reduced as a result of this exercise because nonstrategic buffers (queues) will be eliminated. Again, it is important to remember that, through the continuous improvement process, these buffers will be constantly reviewed and reduced via buffer management.

The last step is the creation of the rope. The function of the rope is to create a communication method to ensure that the release of material into the plant is sufficient to always support the constraint operation. A constraint hour lost now becomes a critical problem. The last thing that is needed is to starve the constraint resource by failing to release material at the proper rate and time. Release is accomplished by communicating the finite production schedule at the constraint back to the first operation supporting that constraint. Since 85 to 90% of all lead time within a job shop is queue and can, therefore, be managed, the communication is simply to eliminate unnecessary lead time from the constraint to the first operation! With only the buffer lead time, it appears that the first work center is always past due. Care must be taken to properly coordinate this change with all management levels involved in the delivery of this material to the constraint so they understand the communication process and are not adversely affected by it. Remember that the lead time causes the WIP inventory in an MRP system as those machine centers in the routing seek to execute the MRP schedule derived from the static lead time calculation. The objective under CM is not to provide a schedule for each machine center but to execute the finite schedule for the constraint. There simply are no schedules printed for machine centers other than the constraint, material release, divergent, convergent, and shipping areas. The gating operation releases material in accordance with the finite schedule at the constraint. Material flows through the shop as required to support the buffers. Material in the constraint buffer is consumed by the constraint.

The execution side requires rigorous adherence to the schedules at the constraints, material release, convergent (assembly), divergent, and shipping areas. The shipping schedule is prepared to support the actual promised ship dates, which are based on the constraint schedule. Purchasing continues to plan its purchase orders that were developed in the MRP system. The shop schedule at the constraint is managed by careful comparison of the actual buffer to the planned buffer. A program can be written to take a picture of the actual WIP inventory position at the buffer on a daily or even an on-line basis.

Using Drum-Buffer-Rope with MRP in a Repetitive Environment

As discussed previously, a significant difference exists between production planning in a job shop and in a repetitive environment. Some modifications in the above application of drum-buffer-rope are required for a repetitive shop. The reason for the change in techniques is the elimination of the queue as both part of the lead time calculation and, most importantly, as part of the actual physical layout of the manufacturing facility. With the elimination of the queue, management is greatly restricted in its ability to alter the plan once a batch of material has been launched. The plan that is created in a repetitive environment is much more likely to be executed as it was developed in spite of any changes that might have occurred.

Envision the machines in a repetitive shop as if there were a single machine cell. This view enhances the understanding of buffer management. A single machine center within the cell paces the production. In many cases, material cannot physically enter the line and stack up in front of the constraint. A transfer line would be an example of this environment. In these shops, several machine lines will probably feed a final assembly line. This configuration is a T-plant, as will be described in Chapter 5. Because a T-plant has a pacing machine center, the scheduling problem becomes one of orchestrating the overall constraint pace from the machine cell to the assembly buffers of the nonconstraint cells in such a way as to allow the execution of the final assembly schedule.

The key modification that is made in the above drum-buffer-rope technique is in the development of the finite schedule at the individual constraints. The first step is to define the capacity of the machine line in terms of the constraint. This is often already accomplished in the development of the MPS, but the process is somewhat the reverse than in a job shop. The management policy for the machine line is determined as one shift, 12 hours, and so on. This policy establishes the upper limit in hours per week. Next, the material requirements from the MRP system are determined for the time period, possibly the quarter or the entire rolling 12 months. The actual cycle time standard of the production requirements is then calculated. This processing time is subtracted from the time available based on the shift manning policy. The residual, which must be positive, is the time available for setups, preventative maintenance, and other planned activities. In some cases, the average downtime is calculated first and subtracted from the shift policy available time. The next task is to develop the proper number of setups that can be made over the given time period. This can be done in a number of ways from the calculation of the contribution margin per constraint minute to the use of a modified economic production quantity by dividing the number of setups allowed into the period requirements. From this production quantity, a run pattern can be calculated, for example, B C B A B A then repeat the sequence. In many repetitive environments, the requirements are spread evenly across the time horizon to allow this method to be used, rather than the choppiness seen in a job shop. The last step in the creation of the finite schedule is to develop the actual production schedule. This is an interactive process that, through trial and error, finds an equilibrium point. From this process, the familiar sawtooth inventory pattern is established. The components then flow into the assembly buffers in such a way as to maintain an adequate supply from which the final assembly schedule can be executed. In like manner, all other machine cell schedules are created based on the constraint schedule and parts supplied into the assembly buffers. Through this process, the drum (the constraint), the assembly, the shipping buffers, and the material release schedules are created.

The material release schedule to each cell represents the rope. Since the lines are likely to prohibit the existence of a large buffer inventory at the constraint, most buffering must occur at the gating operation. If some buffering can be provided prior to the constraint resource and space after the resource (for completed parts), throughput may increase significantly. As no allowance is provided for any queue times, the lead time in a repetitive environment actually reflects the production time to a much greater extent than in a job shop. From the constraint within the machine cell, the actual production times are calculated, and the start times at the constraint are offset to establish the first operation's start time.

Because repetitive shops, by their very nature, overlap operations, the lead times are often much smaller and are typically stated in days or even hours. Through the backward scheduling process the schedules for material release are developed. The constraint buffer is created by one last offset at the first operation by the amount of buffer time that is desired. The purchasing system then takes this gating operation schedule plus new natural buffer into the development of the purchase orders.

In order to execute the finite schedules developed in the repetitive shop, firm planned orders are used in the MRP system. This, it must be noted, also must occur in any MRP application in a repetitive environment in order for valid schedules to be produced.

The advantages of CM can be achieved in an MRP environment. A few modifications to the actual system are required to implement the drum-buffer-rope technique. However, a change in basic management philosophy is also required. The actual systems modifications are the easy part. Without complete understanding and agreement among the management team, this software modification is useless. The last section of this handbook presents the changes that need to be made in the performance measurement systems and the thinking tools used to promote effective problem solving.

CONCLUSION

One final aspect of production planning under the blend of MRP and CM must be presented: the effective use of JIT in the process. While some significant differences exist between the JIT and CM methods, no basic conflict in their objectives exists. JIT attacks the production problem by a rigorous assault on the causes of statistical variability. As a result, over time, a steady flow of material is achieved that greatly eliminates waste from the system. The driver of continuous improvement is inventory reduction throughout the production process.

Under CM, inventory is removed from all operations except where it provides strategic benefits. This CM approach uses inventory to reduce the impact of statistical variability. Throughput and due date performance are protected, while JIT tools can be implemented to eliminate the cause of the variability. Strategic buffering has been likened to the use of sonar to locate and attack the rocks in the frequently seen JIT river analogy. The objective of CM is the same as JIT, however. Both systems seek to eliminate waste from the production process through the use of continuous improvement to achieve the organizational goals.

In the next chapter, a classification framework for manufacturing processes is presented. This general framework, V-A-T analysis, provides the ability to identify the control points across a variety of industries and product configurations.

The five-step focusing process previously presented and the drum-buffer-rope and buffer management elements will be discussed within this framework. The fundamentals can be viewed as an effective production planning and control system.

REFERENCES

1. Goldratt, E.M. and J. Cox, *The Goal: A Process of Continuous Improvement,* 2nd edition. Croton-on-Hudson, NY: North River Press, 1994.
2. Goldratt, E.M., *The Haystack Syndrome: Sifting Information Out of the Data Ocean,* Croton-on-Hudson, NY: North River Press, 1990.
3. Goldratt, E.M., *Production: The TOC Way — A Self-Learning Kit,* New Haven, CT: Avraham Y. Goldratt Institute, 1996.
4. Cox, J.F., J.H. Blackstone, and M.S. Spencer, *APICS Dictionary,* Falls Church, VA: American Production and Inventory Control Society, 1995.

THE V-A-T LOGICAL STRUCTURE ANALYSIS

INTRODUCTION

Because of the traditional way of organizing a business, we tend to view a business as either product centered, comprised of marketing and sales for example, or production centered, comprised of engineers and planners. In fact, the development of computer-based production planning methods like material requirements planning (MRP) reinforced this functional view of a business. This functional view of a business often leads to good managers making bad decisions. In order to eliminate this effect, we need to examine our traditional thinking about an organization.

The V-A-T analysis is an approach that breaks down the traditional barriers and views the organization as an interaction of both products and processes. By seeing the organization in this systems view, three general categories of production structures or shapes emerge, each structure requiring a somewhat different approach to management planning and control. This chapter explains the CM approach, looks at the three general structures, and presents a step-by-step approach to determining and managing a logical product structure.

WHAT IS THE LOGICAL STRUCTURE ANALYSIS?

The first step in taking a systems view of a production process is the development of a logical structure. The logical product structure is comprised of two basic manufacturing documents, the bill of material and the routing. In manufacturing, the bill of material is used to describe the relationship between an end item and its components. Several types of bills of material structures exist. A manufacturing bill of material is used as a guide to actual fabrication of a product from the raw material to the finished end item. More important to the development of a logical structure is the planning bill of material. A planning bill is a summary of product information concerning the relationships among components and end items.

The planning bill does not represent the actual means of assembly, as does a manufacturing bill. The planning bill describes the logic behind the manufacturing bill and is usually used in a computer production planning system such as MRP. The planning bill can contain fictional subassemblies, called phantoms, that are never physically produced but are necessary to force MRP into correctly planning the materials flow. The importance of the planning bill in the development of the logical structure is the understanding that it is necessary to know the underlying relationships supporting the assembly of parts and components into finished goods. These underlying relationships are the basis for the logical structure concept.

A second key component of the logical product structure is the routing. In manufacturing, the routing describes the actual sequence of physical operations that are performed to convert raw material into a usable component. The routing contains information, such as the actual cycle time, standard hours per manufacturing operation, and the machine identification number. As was the case with the bill of material, an underlying logic supports the routing. Routings in some factories may be less rigidly defined than at a machine center located in a manufacturing cell, but the logical flow of material through a series of operations (a manufacturing process) can be used to develop a routing. It is not unusual in manufacturing to have many different routings in the same facility. The number of routings does not impact the overall logical structure.

In V-A-T analysis, the logic contained in the planning bill of material is combined with the logic contained in the routing to create a logical product structure. Surprisingly, three common structures (V, A, T), or shapes, are found when the bills and routings are combined. The name V-A-T originates from the three most common shapes of logical structures. (The I-structure will also be discussed, although other logical structures also exist.) Goldratt[2,3] introduced the V-A-T analysis in his early workshops at Creative Output, Inc. The logical structure is the sequence of operations through which each product (order) must pass in order to manufacture and assemble a product or product family. Using the logical structure concept, a customer order may require the same resource as other customers' orders, and prioritization is therefore required among these orders. For example, suppose all customer orders of a particular type require a packaging operation. The packaging operation will appear in the routings, and therefore the logical structure, for each order. The logical structure will include all operations required to complete an order. This process network concept will be developed more fully in the following sections for each of the logical structures. First, let's examine the simplest structure, an I-structure, to define and illustrate some basic concepts.

WHAT ARE THE BASICS OF LINES?

Traditional line design focuses on balancing the production capacities of all work centers in a production process. It is assumed that if the production capacity of each work center is ten units an hour, then the line output will be ten units an hour. This theoretical output seldom occurs, however. Goldratt recognized this phenomenon as the accumulation of statistical fluctuations across dependent resources and illustrated it in *The Goal*[1] using his scout hike. In *The Goal,* Alex Rogo led a troop of several scouts on a hike. He initially estimated that the troop should average two miles an hour, but after a few hours he noted that the actual pace was far slower. He also noted that as one scout slowed down or stopped, all scouts behind that scout had to slow down or stop. These statistical fluctuations in the pace of each scout were passed along to the whole troop, and the troop could not march any faster than the slowest scout (Herbie). In practice, however, Herbie was impacted by the statistical fluctuations of the other scouts (Herbie also had to stop occasionally based on other scouts), and therefore the troop was actually slower than Herbie. *The Race*[4] also has an excellent discussion of the troup analogy.

Applying these two characteristics, statistical fluctuation and dependent resources, to a simple line leads to the principle in line design of having idle capacity at nonconstraint resources so that they have the capability to catch up with the pace of the constraint. This concept is illustrated in Figure 5.1. Assume that the fourth station in a six-station line is the constraint and has a capacity to produce 10 units an hour; the nonconstraints have a capacity of 12 units an hour. Theoretically, the line capacity is ten units. If breakdowns occur prior to the constraint, the constraint can starve unless enough

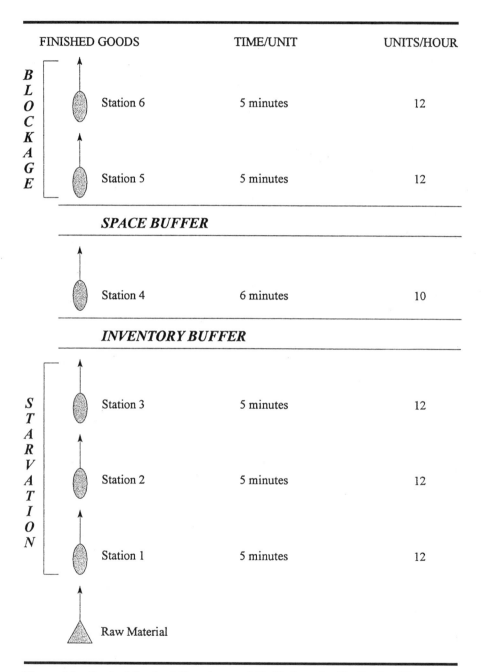

Figure 5.1   Logical Product Structure and Buffering of a Six-Station Line

inventory is located in front of the constraint so that it can keep operating until inventory flow to the constraint is reestablished. If breakdowns occur at work centers after the constraint, then the constraint can be blocked from continuously producing, and therefore line throughput can be reduced. Enough space must be located behind the constraint to off-load parts from the constraint until the station that is down can start again. The buffer in front of the constraint prevents starvation of the constraint and is called an inventory or time buffer. It should be nearly full most of the time. The buffer behind the constraint prevents blockage caused by statistical fluctuation after the constraint and is called a space buffer. It should be empty most of the time. These strategic buffers allow the constraint to operate independently of most problems at other workstations. If these two buffers are sized correctly, then the line throughput should only be impacted by statistical fluctuations (for example, breakage) at the constraint. In constraints management, breakage at the constraint is critical, as it impacts system throughput. Preventive maintenance at the constraint pays bottom-line dividends. By understanding the concepts of starvation, blockage, and breakage and how to reduce the impact of each of them, throughput can be improved significantly at little expense.

The concepts of productive, idle, protective, and excess capacities also are important to an understanding of the design and management of production processes. As mentioned previously, traditional thought is to design and manage a production process so that each workstation has equal capacity. Idle capacity is viewed as an evil and therefore should be eliminated to reduce costs. In constraints management, it is recognized that statistical fluctuation and dependent resources are a reality, and therefore these characteristics should be considered in the design and management of the production process. The definitions of productive, idle, protective, and excess capacity are illustrated in Figure 5.2. The production capability of the constraint is defined as the productive capacity of the line. Any capacity at nonconstraints above this productive level is viewed as idle capacity and in a perfect world (no statistical fluctuations) would be a waste. However, due to statistical fluctuations, some increased level of capacity at nonconstraints is required to prevent nonconstraints from slowing down the constraint pace. This "idle" capacity used to reduce the impact of statistical fluctuations at nonconstraints on the constraint throughput is called "protective" capacity. Any capacity at a nonconstant work center above the productive and protective capacity level is called excess capacity.

An understanding of starvation, breakage, and blockage is fundamental to identifying the causes of lost throughput, and the use of time buffers,

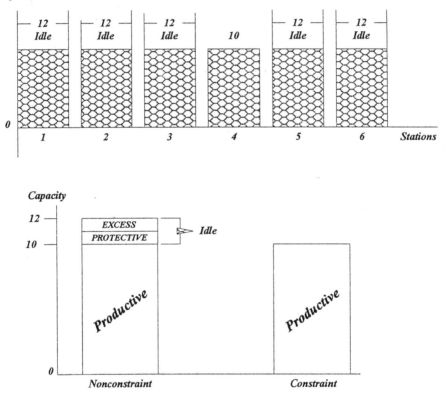

Figure 5.2   The Relationship of Idle, Productive, Protective, and Excess Capacity for Nonconstraint and Constraint Resources in a Line

preventive maintenance (at the constraint), and space buffers provides the solution to maintaining system throughput. In Figure 5.3, the relationship of productive capacity at the constraint to productive and protective capacity at nonconstraints is illustrated. An understanding of these concepts is required to ensure that adequate capacity is provided to determine and maintain system throughput. Both protective capacity and buffers should be used to ensure that constraint throughput capacity equals system throughput. While described in a simple I-structure, these concepts apply to all logical product structures.

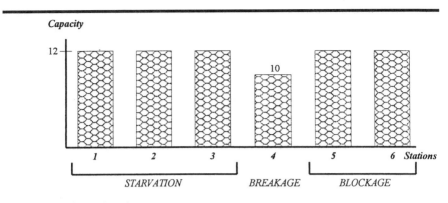

a. Causes of Lost Throughput.

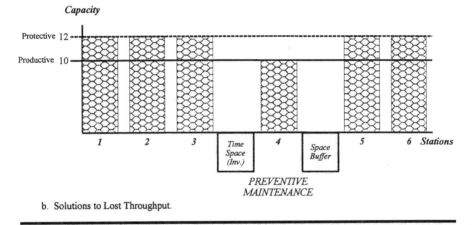

b. Solutions to Lost Throughput.

Figure 5.3   Design Considerations for a Six-Station Line

WHAT IS A T-STRUCTURE?

Figure 5.4 depicts one type of logical structure, the T-shape. This logical structure is the most common in modern production facilities.

As seen in Figure 5.4a, the routing is comprised of a series of seven sequenced steps that result in the creation of a finished product. However, more than one deliverable item that can be created from the same steps. This logical structure describes the flow of common parts and assemblies across the same workstations to produce and/or assemble a variety of similar products

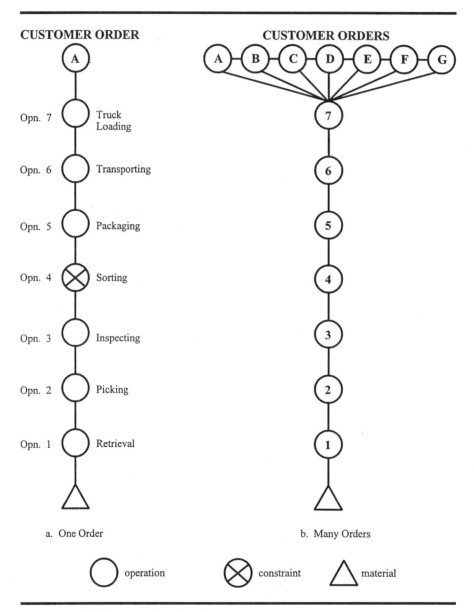

Figure 5.4   T-Shape Logical Structure

and graphically resembles a T-shape (Figure 5.4b). Common characteristics of T-structures create managerial problems and have significant implications for the planning and control systems.

The most common characteristic of a T-shape logical structure is that it is possible to deliver numerous combinations of end products from a limited number of similar steps. Final assembly and delivery are typically based on actual customer orders (make- or assemble-to-order products) rather than a forecast. Often excessive work-in-process and finished goods inventories are maintained as safety stock to ensure meeting customer orders immediately upon receipt. The remaining activities (after receipt of customer order) are usually labor intensive, with picking, assembling, and packing operations, and overtime often used to meet schedules. A key managerial problem is misallocation of a common assembly from one product to another product or part of a shipment from one customer order to another in order to meet schedules. This misallocation causes expediting to fill the original customer order and often results in further misallocations and overtime.

Clearly, the logical structure described in Figure 5.4 is a very simplified version. Figure 5.5 depicts a more realistic logical structure that might be found in a production facility.

In V-A-T analysis, focusing attention on a small number of work centers within the logical structure leads to improved performance. One control point is the constraint operation. Every system has a constraint. In Figure 5.4a, the constraint was identified as the sorting operation (opn. 4). (Under CM, the five-step focusing process discussed in Chapter 3 is used to identify the constraint.) Note that the sorting operation exists in most, but not all, of the logical structure branches. This indicates that most end product orders require sorting. The pace established by the sorting operation determines the number of end items that can be delivered in a day. Other operations are subordinate to the constraining operation pace. As a result, management should ensure that the sorting operation is fully staffed at all times during the working shift. An hour lost at the constraint is an hour lost throughout the system.

Another management focal point is the gating operation, that is, the first operation in the system. For example, if the constraining operation is fully staffed and produces at a rate of ten orders per hour, what will be the effect if more than ten orders per hour were released into the system? The result cannot be additional system output because this output is determined by the constraint, which has not changed. The result will be an increase in the work-in-process inventory levels and the likelihood of confusion among workers

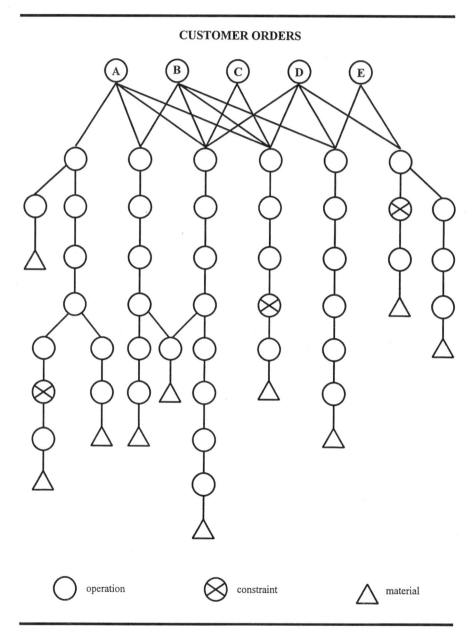

Figure 5.5   Realistic T-Shape Structure

concerning the relative priorities of the various orders. The increase in work-in-process may cause the misallocation of components to occur as one part of an order is combined with other parts of an order for shipment. Essentially, the factory is robbing Peter to pay Paul. Eventually, the orders now missing components will be needed for shipment. Another type of misallocation can occur when labor is assigned to work on an order that is not immediately needed, while other more critical orders await their turn. The misallocation of components or labor to other orders may cause increased expediting. In our example, the sorting operation is still the constraint and operates as before. Overall throughput remains the same. Therefore, to avoid confusion, reduce the opportunity for misallocation, and reduce expediting, the number of orders released into the warehouse system should not exceed the amount of production that occurs at the system's constraint, the sorting operation. Management controls that number of orders released to the production system at the gating operation and, thereby, improves the overall productivity.

Management is frequently concerned with employee utilization in a factory. As seen in Figure 5.4, several nonconstraint operations are required. Management has a natural tendency to insist that all employees (even at nonconstraints) are productive all of the time. However, if the previous discussion concerning the constraining operation is valid, then no additional orders (above the pace of the constraint) should be released into the factory. As a result, nonconstraint operations will, by definition, be underutilized some of the time. The orders that do not require a constraint operation (no sorting in this example) can be released to the factory based on the market demand without creating a backlog of work at the constraint. Care must be exercised not to create interactive constraints.

Because management now can focus attention on the constraining operation, workers at nonconstraints can simply work on orders as material is released into the factory based on actual customer demand. Actual lead times for orders should, therefore, be reduced because workers at nonconstraints should only be working on actual customer orders. The system should be able to respond to short-run changes because there will be less work in process overall. Less work in process reduces the opportunity for misallocation and the need for expediting.

The constraint operation establishes the overall rate of system throughput. Management must now take action to ensure that the constraining operation is always active. This is accomplished by the time buffer in front of the constraint. An error that occurs at a work center which precedes the constraining operation can "starve" the constraint. In Figure 5.4, an error in the picking operation (opn. 2) can occur and will cause the sorting operation to be idle while the correct parts are picked. Any time lost at the constraint is lost to the system, however. Therefore, in order to protect the constraining operation from any of the normal instabilities (statistical fluctuations) that can occur in an actual factory, a strategic buffer of work is located just prior to the constraint.

The constraint determines the system throughput. A finite schedule of the sorting operation (the constraint) is developed based on the order backlog. This backlog establishes the priority of orders at the constraint. An additional benefit of this approach is that the employee(s) at the constraint now has access to the actual priority of final customer orders. The possibility of miscommunication of order priorities is reduced or eliminated.

The production out of the constraint matches final customer orders. The packaging, transporting, and truck-loading operations (the succeeding operations from the constraint) are usually performed by employees in the order that emerges from the constraint. Therefore, the actual order priorities are maintained. Since neither of the succeeding operations is a constraint, protective capacity, usually in the form of employee time, is available to meet delivery commitments. Any expediting that may be required can be accomplished by using the employees assigned to nonconstraint operations, as was discussed in previous paragraphs.

As a result of the use of the V-A-T logical structure analysis, management is able to focus on relatively few control points in a production system. By creating a relatively small buffer the constraining operation can be utilized to maximize throughput. Customer demand is enhanced as misallocations, the major problem in a T-structure, are eliminated.

WHAT IS A V-STRUCTURE?

The second common type of logical structure is a V-shape. Figure 5.6 depicts a V-shape logical structure. The characteristics of a V-shape logical structure differ from a T-shape logical structure. The principal difference is that, unlike a T-structure, there are very few, often only one, types of material that are used to supply the customer base. This single type of material, however, is used to provide a wide variety of unique products to customers. Once past an operation, the material generally cannot be rerouted to a different product. The cooking of a steak rare, medium, or well done represents a divergent

CUSTOMER ORDERS

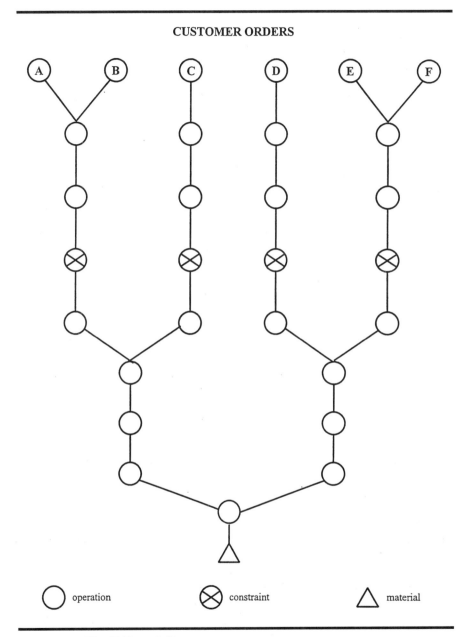

Figure 5.6 The V-Shaped Structure

point in the V-logical structure of preparing a steak dinner. Once the steak has been cooked well done, it is impossible to reroute it to convert it to a rare steak. The customer order often requires customization. As a result of the uniqueness, opportunity exists for misallocation of the product to another product. The entire order quantity may be diverted to another customer.

A V-shape logical structure usually contains a divergent operation early in the logical structure that prevents the shifting of an order except possibly through a costly reclaim procedure. Several points of divergence can exist within the logical structure that are "points of no return."

The divergent control point dominates the V-shape structure, while the convergent control point dominates the T-shape logical structure. The management of these control points is similar. As was the case in a T-shape structure, a constraint sets the system throughput pace. In many cases in a V-structure, the constraint is located at a divergent operation. The constraint is managed in the same manner as was previously discussed.

A buffer is created to protect the constraint from any disruptions from the preceding activities. A space buffer is placed after the constraint to protect against disruptions at downstream operations. The constraint is used to pace the orders released into the system, as was the case with the T-shape structure. The customer orders are used by management to create a finite schedule for the constraining operation. Orders may be scheduled at the constraint to reduce setups. Finally, the constraint is assigned employees to maintain its production pace throughout the scheduled shifts.

The gating activity in the V-structure is also managed in the same manner as in the T-structure. Management of material release is usually easy in a V-structure. Generally one gating operation exists in a V-structure, whereas in a T-structure several gating operations exist. Unless an order does not require the constraining activity, which is rare in a V-structure, the orders are released at the same rate as orders are produced at the constraint.

The divergent point must be planned and controlled carefully in the V-structure. Management must control the allocation of material to final customer orders at the point(s) of divergence. Work centers often misallocate materials to reduce setup time and increase department efficiencies. There is generally no need for a buffer to exist at a nonconstraint divergent point, because some protective capacity is available to correct errors from preceding activities and allow ample time for setups. Divergent points operations cannot be measured by departmental efficiencies. Control of the allocation of material at the divergent point prevents order from being misdirected to a different customer. The divergent point is controlled by a schedule that is derived from both the constraint schedule and the final customer order. The employee at the point of divergence is provided the information concerning the customer order priorities and quantities. In rare cases, the divergent schedule may have to be offset from the constraint schedule because of the lead times. In warehousing, the lead times are rarely long enough to require an offset procedure. If more than one divergent point is contained in the structure, then management supplies each divergent point with a prioritized schedule. In process industries, long setups which create near constraints may occasionally exist. In this situation, a buffer may be inserted at the divergent point to combine orders into batches to reduce setup time and thus reduce the impact of the near constraint.

WHAT IS AN A-STRUCTURE?

Figure 5.7 depicts the third common type of logical structure. The A-structure is characterized by having numerous combinations of activities that are required to provide relatively few end items for customer orders. The A-structure usually has a wide variety of resources that are used to create the customer orders. As was the case with a T-shape structure, there are points where different components or subassemblies converge to complete a single order. However, in an A-structure, there are activities that occur after the convergence of the resources, and the A-structure tends to have fewer common components and tends to supply fewer end items to final customers than the T-structure. Another unique characteristic of an A-structure is the wide variety of routings, or sequence of activities, that are required to fulfill customer orders. Both the V-structure and the T-structure have relatively few routings to manage. In the A-structure, each end customer order may require a unique series of activities that may never be repeated. Employees working in an A-structure environment (often a job shop) tend to be generalists in their skills and work assignments. There is usually a significant amount of expediting as workers are assigned and reassigned throughout the day as customer priorities change or as material availability changes.

Common control points used to manage a T-structure are also are used to manage the A-structure. Even though a unique series of activities may be required for many end customer orders, a constraining operation usually exists within the facility. In an A-structure, the constraint is more difficult to identify, however. The constraint may be a particular employee skill or a

CUSTOMER ORDERS

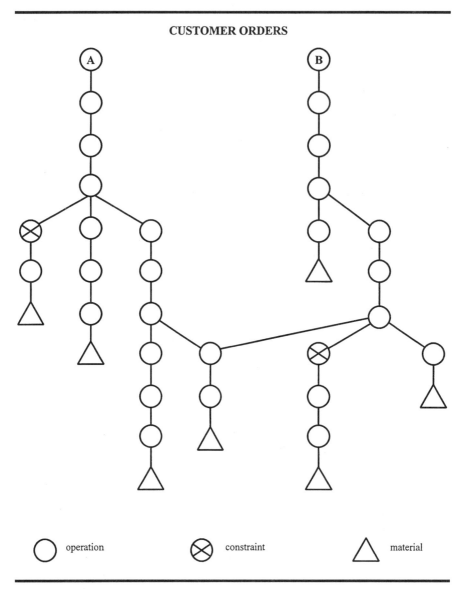

Figure 5.7   The A-Shape Structure

piece of equipment that is required for almost all orders. The constraint is managed in the same manner as was previously discussed. A buffer is created in front of the constraint, and a finite schedule is developed based on customer orders. A space buffer is also planned behind the constraint. As before, the constraint establishes the pace of order delivery, and materials for new orders are released in accordance with the production completed at the constraint.

Like the T-structure, the primary control point in an A-structure is the convergent or assembly point. The difference is in the relative position within the logical structure of the convergence points. In the T-structure, the convergent points are at assembly and packaging and cause subassemblies to be misallocated to a specific customer order. In the A-structure, the convergent points are generally fabrication areas and cause misallocation of capacity from one part to another.

The management of the convergence points is the same as in the T-structure. That is, a finite assembly schedule is prepared for nonconstraint parts that are to be assembled with constraint parts. This schedule is based on the constraint schedule and customer order priorities and insures that nonconstraint parts arrive at assembly points prior to constraint parts. The same method is used to control any divergent points in an A-structure as well. The additional schedule provides employees with the actual quantity that is required at the time required to complete a customer order. As a result the employee is far less likely to misallocate capacity to the wrong order, and expediting is greatly reduced since the flow of material is controlled by the assembly schedule.

THE DIFFERENCE BETWEEN V-A-T AND JIT STRATEGIC INVENTORY MANAGEMENT

The lessons learned from the Just-in-Time (JIT) experience should not be lost. The misuse of inventory buffers is a waste of resources and retards a firm's ability to compete. Zero inventory is a JIT goal. However, a practical concern for managers is the effective operation of the production facility until JIT can be completely implemented. (This includes all of the necessary prerequisites, including setup reductions, quality at the source, and a truly flexible work force.) JIT implementation requires time and dedicated effort throughout the organization. Before the JIT ideal of minimum inventories can be realized, the use of buffer management via V-A-T logical structures has positive implications in the day-to-day operation of the production environment. The application of constraints management provides a powerful management tool during the transition. Further, the use of strategic buffers can facilitate the transition to the JIT ideal by permitting an increase in throughput and a more effective production environment, thereby allowing management to focus on the JIT prerequisites.

An interesting problem emerges in migrating to the idealized JIT environment. The law of diminishing returns appears to apply. Suppose we implement a JIT program. Being good managers, we elect to make those changes to the system that have the greatest impact at the least cost. Let's say we engage in a total quality management program and implement control charts. The cost is relatively small compared to the benefits of eliminating scrap. Next we revise a setup. This costs some amount of money and, by definition, the benefits are less than the first change. However, because the benefits outweigh the costs, we make the change. Now we are faced with making an additional change, perhaps redesigning the layout to provide a U-shape flow. What if the costs are higher than the benefits? Isn't there likely to be a point in the implementation of JIT, or any other program, at which incremental costs exceed the incremental benefits? Should we continue? How such a decision is handled under constraints management will be discussed in Section III of this handbook, but the law of diminishing returns seems to be in effect. It is unlikely, therefore, that a JIT program *can be,* not *should be,* fully implemented as is theoretically possible in an existing factory.

STRATEGIC OPERATIONS MANAGEMENT USING V-A-T ANALYSIS

The strategic use of inventory in constraint, assembly, and shipping buffers ensures the smooth flow of customer orders through the production process. Despite the large number of operations in a modern production facility, management can plan and control a few points and thereby increase overall production. One criticism of the JIT approach has been that it spreads management attention throughout the plant, and as random problems are encountered by lowering inventory, efforts are directed at improving operations and eliminating the obstacles. No guarantee exists that the problems surfaced are those that are the true obstacles actually constraining the plant's throughput. They may be important and worthy of management and employee efforts directed toward their removal, but they may not be the most valuable objectives at the time.

The next most critical parts that support the strategic use of inventory are those that flow into the nonconstraint buffers. These buffers are the convergence points in A-plants and T-plants and those at shipping in all plants.

While not as critical as constraint buffers, their overall accuracy is important in the management of a firm's due date performance and lead times. The function of the buffers is to absorb the unplanned delays that are part of the production environment. By focusing attention on these buffers, the problems which adversely impact throughput timing can be minimized. Like JIT's process of continuous improvement, V-A-T analysis can be used to identify and eliminate the core problems that disrupt system throughput and due date performance.

HOW TO USE THE V-A-T CONTROL POINTS TO MANAGE PRODUCTION

Managerial control under a constraints management system consists of three components: a framework used to identify control points called V-A-T analysis, a scheduling component called drum-buffer-rope, and a problem prioritization component called buffer management (the latter two are discussed in Chapter 4). In this section, the management control of each of the three structures is discussed as we tie all of the components of constraints management together, using Bob's Bolt Company as an example. Let's look at the easiest example first.

The T Logical Structure

In a T-structure, a series of common components, subassemblies, and assemblies (each with its own routing) are combined to create a finished product (see Figure 5.8a). Notice, for example, product X2, which is composed of assemblies AR6 and AR7. More than one item (X1, X3, X4, and X5) can be created from the same common assemblies. When the assemblies are combined with the logic that multiple products can be created, a T-shape emerges. Significant implications result in the planning and control of the process, however. The T-shape has several common characteristics which cause managerial problems.

The most common characteristic of a T-shape logical structure is that numerous combinations of end products can be delivered from a limited number of common subassemblies or assemblies. Final assembly of products is typically based on actual customer orders rather than a forecast. Often, excessive work-in-process inventories are maintained as safety stock to ensure meeting customer orders immediately upon receipt. Additionally,

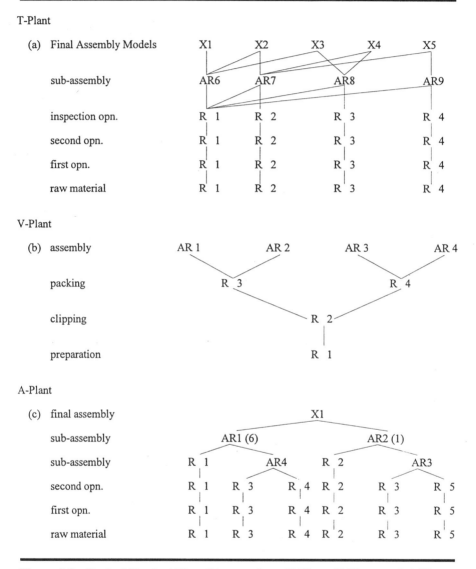

Figure 5.8   Typical Product Flow Diagram for a T-Plant, V-Plant, and A-Plant

assemblies scheduled for one customer order are used to complete another customer order. The assembly activities tend to be labor intensive, and over-time is commonly used to meet schedules. A key managerial problem is the misallocation of a common part or component of one product to another product to meet schedules. This misallocation causes expediting of replacement parts to assemble the product to fill the original customer order and often results in further misallocations and overtime in fabrication areas to produce missing components.

V-A-T analysis focuses management attention on a relatively few critical operations, providing effective management of the entire process. One control point is obviously the constraint which exists in every system. Let's assume that the constraint in Figure 5.8a was found to be the second R2 operation. Under constraints management, the five-step focusing process is used to manage the constraint. The pace established by this operation determines the number of end items that can be delivered in a day. Other work centers are subordinate to the constraining work center. As a result, management should ensure that R2 is fully staffed at all times during the work shift. An hour lost at the constraint is an hour lost throughout the system. The constraint is the drum, the buffer is a physical amount of inventory strategically positioned to keep the constraint operating by absorbing the normal instabilities in the production process prior to the constraint, and the rope is a communication devise from the constraint to the gating operations to trigger release of additional material to the shop floor.

Another management control point is the gating operation, that is, the first operation in the system. For example, if the constraining operation is fully staffed and produces at a rate of ten units per hour, what will the effect be if more than ten units per hour are released into the system? As we found before, the result cannot be additional throughput because the throughput is determined by the production rate of the constraint, which has not changed. The result will be an increase in the work-in-process inventory levels, which increases the likelihood of confusion among workers concerning the relative priorities of the various orders. The increase in work-in-process may cause the misallocation of components to occur, as one part of an order is combined with other parts of an order for shipment. Simultaneously, capacity at nonconstraints is misallocated and can result in creating temporary constraints. Eventually, the orders now missing components will be needed for shipment. The second type of misallocation can occur when capacity is assigned to work on an order that is not needed immediately while other more critical orders await their turn. The misallocation of components or capacity to other orders may cause increased overtime and expediting. In our example, the R2 operation is still the constraint and operates as before. Overall throughput remains the same. Therefore, to avoid confusion, reduce the opportunity for misallocation, and reduce expediting, the number of orders released into the system should not exceed the amount of production that occurs at the system's constraint. Management controls the amount of orders released to the system at the gating operation and thereby improves overall productivity.

Remember that management is frequently concerned with employee utilization in a factory. As seen in Figure 5.8a, several nonconstraint operations are required. A natural tendency is for management to insist that all employees be productive all of the time. However, if the previous discussion concerning the constraining operation is valid, then no additional materials should be released for production. As a result, nonconstraint operations will, by definition, be underutilized some of the time. The orders that do not require a constraint operation can be released to the factory without creating a backlog of work at the constraint, but these orders should be released based on market demand.

Because management now can focus attention on the constraining operation, workers at nonconstraints, where there is excess capacity by definition, can simply work on parts as material is released into the factory to support actual customer demand. Lot sizes can be reduced at nonconstraints, thus reducing nonconstraint idle time. Actual lead times for orders should be reduced further because there will be an increase in worker flexibility (at nonconstraints) to respond to short-run changes and there will be less work in process overall. Less work in process reduces the opportunity for misallocation and the need for expediting.

Because the constraining operation establishes the overall rate of output for the system, management must now take action to ensure the constraining operation is always active. An error that occurs at an activity that precedes the constraining operation can "starve" the constraint. In Figure 5.8a, an error in a preceding work center can occur and will cause the R2 operation to be idle until parts are expedited to R2. Any time lost at the constraint is lost to the system, however. Therefore, in order to protect the constraining operation from any of the normal instabilities that can occur in an actual production facility, a buffer of work is located prior to the constraint.

The production of the constraint matches final customer orders. The succeeding operations after the constraint are usually performed by employees in the order that work emerges from the constraint. Therefore, the actual order priorities are maintained. Because the succeeding operations are nonconstraints, protective capacity, usually in the form of employee time, is available to meet delivery commitments. Any expediting that may be required can be accomplished by using the employees assigned to nonconstraint operations.

V-A-T logical structure analysis focuses management's attention on a few control points in a production system. By creating a small buffer prior to the constraining operation, throughput is protected and maximized. Customer demand is enhanced as material misallocations at assemblies and capacity misallocations in fabrication, characteristic of a T-shape, are eliminated.

Now let's look at a more complicated structure, a V-structure, that exists at Bob's Bolt Company.

The V Logical Structure

A V-shape logical structure is presented in Figure 5.8b. Some characteristics of a V-shape logical structure differ from a T-shape logical structure. The principal difference is that, unlike a T-structure, very few, often only one, materials are used to make the product. This single material, however, is used to provide a wide variety of unique products to customers. Once past a divergent point in its routing, the material is generally unable to be used to fill any other order because it often requires customization. Usually during fabrication, the entire order quantity is diverted from the desired product to another product. This misallocation of material is the result of a department's desire to reduce setup times and increase its efficiency measure. Frequently, points of divergence within the logical structure are "points of no return." When this occurs, new material must be released and expedited to replace the order, and frequently numerous setups must be broken to meet the customer's desired due date.

The control points for a V-shape logical structure are different, although the management of the control points is similar. As was the case in a T-shape structure, a constraining operation sets the pace of output for the entire system. The constraint is managed in the same manner as was previously discussed.

A constraint buffer is created to protect the constraint from any disruptions from the preceding activities. The orders released into the system are based on the constraint's pace, as was the case with the T-shape structure. The customer orders are used by management to create a finite schedule for the constraining operation. Orders are frequently sequenced to reduce setups at the constraint. Finally, the constraint is assigned employees to maintain its production pace throughout the scheduled shifts.

The gating activity in the V-structure is also managed in the same manner as in the T-structure. Unless there is an order that does not require the constraining activity, which is rare in a V-structure, the orders are released at the same rate as orders are produced at the constraint.

There is, however, an additional control point in the V-structure. Management must control the allocation of material to final customer orders at the point(s) of divergence. Usually no need for a second buffer exists at the divergent point since, as a nonconstraint, some protective capacity is available to correct for delays from preceding activities. Control of the allocation of material at the divergent point must be exercised in order to prevent the misallocation of the entire order to a different product. The divergent point is controlled by using a second finite schedule that is derived from customer orders. This schedule links the constraint schedule to the shipping schedule. The employee at the point of divergence is provided information concerning the customer order priorities. In rare cases, the divergent schedule may have to be offset from the constraint schedule because of the lead times. If there is more than one divergent point, then management supplies each point with a schedule of order priorities.

Now let's look at the most complicated of logical structures, the A-structure.

The A Logical Structure

The third major logical structure, the A-structure is diagrammed in Figure 5.8c. The A-structure is characterized by having different routings for numerous parts that converge to a relatively few end items. The A-structure usually uses a wide variety of resources to create these end products. As was the case with a T-shape structure, there are points where different materials converge to complete a single product. However, in an A-structure, activities may occur after the convergence of the parts into assemblies, and the A-structure supplies fewer end items to final customers than the T-structure. Another unique characteristic of an A-structure is the wide variety of different routings required to produce end items. Both the V-structure and the T-structure are comprised of relatively few routings. In the A-structure, a few end items consisting of unique assemblies/subassemblies are comprised of many different parts with different routings. Employees in an A-structure tend to be generalists in their skills and work assignments. There is usually a significant amount of expediting as operations are assigned and reassigned throughout the day as customer priorities change or as material availability changes. Some A-structures may evolve into T-structures as more product features and options are offered.

Some control points used to manage an A-structure are the same as those used to control the T-structure. In an A-structure, the constraint is more difficult to uncover. The constraint may be a particular employee skill or a piece of equipment that is required for most orders. The constraint is managed in the same manner as was previously discussed. A buffer is created in front of the constraint, and a finite schedule is developed based on customer orders. The constraint establishes the pace of order delivery, and materials for new orders are released at gating operations in accordance with the production completed at the constraint.

Another control point in an A-structure is a convergent point. A convergent point is a point in the logical structure where two or more parts/components are combined into a subassembly. To complete the assembly operation, all components must be available. An assembly buffer is used to monitor the progress of parts/components to the assembly work center. Misallocation of capacity at a proceeding fabrication work center caused by batching jobs to save setups usually causes tardiness at the assembly area. The assembly buffer is used to maintain part priorities and flow into the convergent points.

The control of the convergent points is similar to the control of the divergent points. That is, a finite schedule is prepared for the assembly buffer based on a finite schedule of the customer orders used at the constraint. The additional schedule provides employees with the actual quantity that is required at the time required to complete a customer order. As a result of the schedule, an employee is far less likely to misallocate capacity to the wrong order, and expediting is greatly reduced because the flow of material is controlled by the constraining activity.

In constraints management, five control points are used to manage a production process. The shape of the logical structure (V, A, or T) influences the dominant control points in the process. The first control point is the constraint. The second control point is the points of divergence, if any. The third control point is the points of convergence, if any. The fourth control point is the gating operation. The fifth control point is the shipping operation. The constraint determines the throughput for the entire system and is controlled by the drum-buffer-rope scheduling method. The constraint schedule, as depicted in the buffer, determines the supporting schedules for planning and controlling the gating operation, shipping operation, and points of divergence and convergence. The development and use of the buffers and the supporting schedules are presented in the following section.

HOW TO IMPROVE PRODUCTION PERFORMANCE

By using the V-A-T logical structure analysis, by viewing a production process as a system, and by planning and controlling material flows using the control points, significant improvements in production processes can be made. The management of a factory is not a uniform set of activities but should be a function of the overall process. Different factories have different processes and require different management. Planning and controlling using the control points and buffer management give managers the ability to better focus management improvement activities. The control points in each logical structure are identified in Figure 5.9.

The productivity improvements that have resulted from the application of V-A-T analysis, in both manufacturing and service applications, emphasize the importance of viewing production as a process rather than as isolated operations. By viewing the overall process, the nature of the limiting constraints can be isolated and improvements can be focused where they can do the most good. Warehousing has come a long way from being viewed as simply the storage of inventory for just-in-case use. As productivity gains are made and excess inventory is eliminated, the role of the factory as an integral part of the overall logistics channel is gaining acceptance. The overall logistics channel itself can be viewed as a production process, and efforts can be directed toward improving the entire logistics process rather than viewing each channel element in isolation.

THE CONTINUOUS IMPROVEMENT PROCESS USING CONSTRAINTS MANAGEMENT VIA CASE STUDIES

We are now in a position to use our knowledge of constraints management to understand how it is used in organizations. The next section of this handbook presents three cases where constraints management has been used to successfully manage a factory. In Chapter 6, an overview of the cases is provided, showing how they fit into the framework developed in Chapter 2. The chapter also includes a discussion of how the case studies were developed, how to use the case studies to expand your understanding of constraints management, and how you can develop a similar analysis of your own factory.

Chapter 7 presents a case study of Velmont Industries, a V-structure. All of the various control points and the use of drum-buffer-rope are discussed.

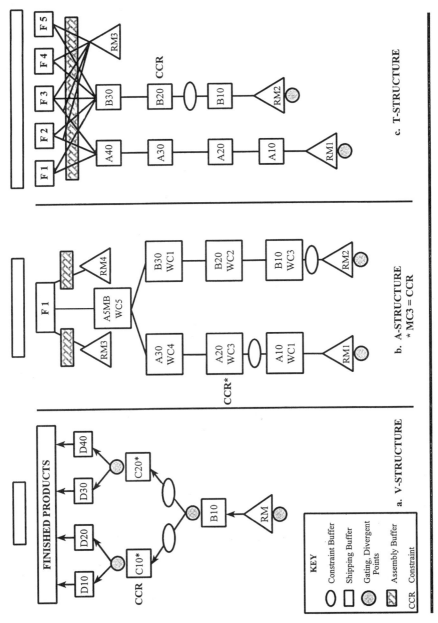

Figure 5.9   V-A-T Logical Structure Control Points

In Chapter 8, the Trane Company, a T-structure, is examined. In Chapter 9, a more complex case involving the Stanley Furniture Company is presented. Stanley consists of seven separate plants that interact to form a manufacturing complex. The five plants that are A-structures are discussed first, followed by the two plants that are V-structures. The focus is on how the plants interact under constraints management to manufacture a complete array of products.

REFERENCES

1. Goldratt, E.M. and J. Cox, *The Goal: A Process of Continuous Improvement,* 2nd edition. Croton-on-Hudson, NY: North River Press, 1994.
2. Goldratt, E.M., "Gaining Momentum with Focused Productivity Improvement" Workshop. Milford. CT: Creative Output, Inc., 1987.
3. Goldratt, E.M., "Staying Ahead with a Process of Ongoing Improvement" Workshop. Milford. CT: Creative Output, Inc., 1987.
4. Goldratt, E.M. and Fox, R.E., *The Race*, Croton-on-Hudson, NY: North River Press, 1986.

CASE STUDIES OF CONSTRAINTS MANAGEMENT FACILITIES

II

AN OVERVIEW OF THE CASES AND ANALYSIS

INTRODUCTION

This chapter describes the case study methodology and provides an overview of the case studies presented in the next three chapters to illustrate the implementation and use of constraints management methods. This chapter can be omitted if one is interested only in the practical aspects of constraints management implementation and use, but it is important for researchers and others interested in developing case studies for their own organizations.

The next three chapters examine actual companies that have implemented constraints management. The cases can be used separately but actually build upon one another from a company with a fairly simple logical structure to a company that consists of plants with complex interacting structures. The first case study, presented in Chapter 7, is Velmont Industries, an example of a V-structure. The second case, presented in Chapter 8, is the Trane Company's Macon, Georgia factory, an example of a T-structure. In Chapter 9, the Stanley Furniture Company's plants that are both A-structures and V-structures are examined. The Stanley V-structure plants interact among the Stanley A-structure plants to provide a fairly complex organization that uses constraints management.

In order to help the reader refer back and forth among the case studies, the same general format is used throughout the cases. The general case format is as follows:

I. Key characteristics of the company
II. Description of the production planning and control methods

   III. Key problem areas identified by management

   IV. A case summary

   V. A within-case analysis discussing key factors

Each of the cases follows the same format in order to facilitate cross-referencing and consistency. First, the key company characteristics are presented, including the company background, the physical operations, the production planning environment, the organizational structure, and the product-process logical structure. Second, the methods used to manage the five production planning and control functions (master production scheduling, priority planning, capacity planning, priority control, and capacity control) are discussed. Third, the key problem areas in each of the five production planning functions are presented. Fourth, a brief summary is presented. Finally, a within-case analysis is presented which describes the relationships and critical characteristics found in each company.

Additionally, a series of questions is provided which the reader can answer in terms of his or her own organization (Figure 6.1). By following the outline provided above and answering these specific questions, the reader can develop his or her own company analysis from which comparisons to the cases included in this handbook can be made. The reader then can use the key problems section to uncover similar problems in his or her own organization. Also, the within-case analysis section for the appropriate logical structure can be used to highlight areas where key factors interact within the production planning and control system to determine overall success.

RESEARCH METHODOLOGY AND CASE STUDY DEVELOPMENT

As indicated in Chapter 2, little research has been conducted on constraints management production planning and control systems. The major research involving the five functions has been the case studies developed by practitioners. Academic articles involved comparisons of systems in a theoretical framework or articles which assumed an idealized state. Practitioners have identified key characteristics which they report require significant changes in traditional material requirements planning and just-in-time systems. Because of these research gaps, this particular research is, by its nature, exploratory and has used the case study methods of qualitative research. The following paragraphs provide the foundation for using case studies to develop the conclusions offered in this handbook.

Marshall and Rossman[1] identify purposes for case study: "Case study research has several purposes—to chronicle events; to render, depict, or characterize; to instruct; and to try out, prove or test" (p. 44). They further identify the applicability of case research: "to investigate little understood phenomena, to identify/discover important variables, to generate hypotheses for further research" (p. 78). Research questions would include: "What are the salient themes, patterns, categories in participant's meaning structures? How are these patterns linked with one another? How do these forces interact to result in the phenomenon" (p. 78)? Patten[2] also identifies areas where the case method is valuable. "Case studies are particularly valuable when the evaluation aims to capture individual differences or unique variations from one program setting to another, or from one program experience to another" (p. 54). Babbie[3] also identifies areas where field research is appropriate. He states, "...field research is especially appropriate to the study of those topics for which attitudes and behaviors can best be understood in their natural setting" (p. 262). Meredith et al.[4] discuss the applicability and need for alternative research paradigms especially in operations management, as do Flynn et al.,[5] who stress empirical research methods such as case studies.

This book follows the Marshall and Rossman[1] framework in developing qualitative research and the framework discussed by Eisenhardt[6] for qualitative research. Site selection was purposeful. Companies volunteered to be studied, had constraints management production planning and control systems and were V-, A-, and T-structures. Various experts from APICS were contacted to recommend appropriate sites. From these sources, a list of potential sites was developed and initial contact made with the factory management. A previsit survey (see Figure 6.1) was administered to potential sites to improve the likelihood of both research fit and researcher acceptance. Once agreement was reached to participate in the study, arrangements were made for an on-site interview.

DATA COLLECTION

An on-site interview instrument was developed and tested to act as a plan for guided discussions with participants (see Figure 6.2). After an extensive interview with the key personnel, triangulation activities were undertaken. These triangulation activities included an on-site plant inspection; interviews with line employees, supervisors, and other functional managers; and the collection of planning and control documents. A set of notes was developed by the researcher during the interview process based on the on-site instrument. Where possible, additional notes were taken during the triangulation process. If permission was given and conditions permitted, a tape recording was made of that portion of the on-site instrument that specifically dealt with characteristics of the five production planning and control functions.

A case study was prepared following the outline of the on-site instrument. This case study was sent back to the research site, and the key participant was asked to review and verify the data. The key participant was also asked to review the case with any others who were involved in the study. Finally, permission was requested to allow publication of the case with appropriate identification of the organization.

DATA ANALYSIS

Detailed case studies were prepared in the manner described in the previous section. From these case studies, information was gathered to answer the basic questions. Case study construction followed the recommendations described by Strauss[7] (pp. 218–240) and Patton[2] (pp. 384–400). Data from the cases were coded, and key characteristics were identified by a cross-comparison analysis.

A detailed within-case analysis was developed for each of the companies. Each within-case analysis consists of a brief introduction followed by a presentation of the key characteristics and the impacts of the key characteristics upon the five production planning and control functions. The key characteristics were identified based on the on-site interviews and the previsit questionnaire. The impacts of the key characteristics of the planning and control systems were also identified from the interviews.

LIMITATIONS

As with any research endeavor, there are limitations and cautions to be exercised concerning the conclusions reached from the study. In this section, readers are warned of the hazards of basing conclusions on case studies.

Validity and Reliability

Huck[8] (Chapter 11) identified seven threats to internal validity. The historical threat was closely monitored by inquiring and noting any key changes concerning the business environment during the previous six-month time period. The threat of instrumentation was monitored by closely adhering to the on-site instrument. There was likely a slight variation from case to case as the researcher gained experience in administering the instrument. A set of field notes was prepared immediately after the on-site administration to provide the researcher with the opportunity to note any concerns. Selection is the greatest threat to validity. It must be noted that the researcher provided no input into the management decisions concerning the selection or application of the production planning and control systems under study. However, it is possible that the selection of one company versus another could alter the findings. A previsit questionnaire was used to screen candidates to reduce this threat. Also, as a check on this problem, considerable information is included in the case itself to address system selection and operation issues. Finally, information is presented concerning the identity of the companies involved in the study for use by other researchers. The threats of maturation, testing, mortality, and statistical regression were controlled in the experimental design.

Huck[8] (Chapter 12) also identified 11 external threats to validity. Population threats are inherent to qualitative research. The selection of site cases was not random, and no expectation of statistical inference is expected. That is the nature of exploratory research. Ecological threats concern problems in the description of variables. To a degree, this was monitored by adhering to the definitions identified in the theoretical development section. Other ecological threats come from the interactions of history, time, and multiple effects. The nature of production planning and control in a real setting is such as to expect interactions to occur. The researcher attempted to note interactions in the case history as they were identified. It should be noted that the actual production environment is one of constant interactions in order to accomplish business objectives. The robustness of production planning and control systems is sufficient to routinely allow these interactions. Since all three systems were studied in actual operating environments, that amount of interaction could be expected to be approximately equal among those systems and, therefore, less of a threat than under true experimental conditions. The last external threats to validity involve the conduct of the participants relative to their conduct in a nonresearch environment. The Hawthorne effects, novelty and disruption, and the Rosenthal effect were all possible. However, because the actual observations were limited and involved studying an ongoing operation, any effects on the production planning and control systems would be expected to be minimal. Experimental reliability was enhanced by the use of a previsit survey and on-site interview instrument, both of which are included. The identity of the sites is given if possible.

An additional threat to both validity and reliability concerns the information received from the site sources. Sources could give misleading information. Sources could give information that they believe is correct but, in fact, is incorrect. Finally, sources could give correct information but the researcher could report it as false information. These threats are controlled, to a degree, by the researchers' background in operations management and production planning and control system design and operation. Companies in the study were rechecked for verification and clarification if these threats occurred.

Generalizability

Eisenhardt[6] identifies generalizability as a weakness in case study theory-building. With only three companies to be studied, generalizability is a concern.

Finally, even though it was possible to develop some conclusions from this research, the passage of time and the increase in practitioner skill and knowledge about the operation of production planning and control systems could reduce the overall utility of the conclusions. Breakthroughs in operations management are possible, and the critical nature of the subject under study make such an event possible. Time and education may work to reduce some of the effectiveness of the conclusions.

HOW TO IMPLEMENT CONSTRAINTS MANAGEMENT

A first step in implementing constraints management is to analyze your own company. What is the current state of affairs? This current state will serve as a benchmark against which to measure your progress.

Your Own Case Study Questionnaire

The questions used by the authors to develop the case studies are provided in Figures 6.1 and 6.2. By answering the questions and using the general outline discussed earlier, the reader can develop a case study of his or her own organization. A few comments might be helpful. These questions were developed so that they could not all be answered by a single person. Several people within an organization will have to be questioned. In this way, there is a check and balance built into the methodology. In qualitative research, this is called triangulation, and it is important in keeping the case study valid.

Also, as many copies as possible of reports and in-house documents should be obtained. This will help clarify some points and serve as a very good reminder about some aspects of the organization that might not be clear. In addition, a diagram of the facility layout should be prepared to facilitate your understanding of the flow of material. While it is unlikely that a logical structure diagram can be developed for the entire product line, by using the above questions and the major product line as an example, you can develop a good understanding of what type of logical structure—V, A, or T—or a combination of the three best describes your company.

Once you have determined the organization's logical structure, you can use the five-step focusing process, along with your case study, to determine the appropriate managerial control points. The next step is to develop a plan to implement drum-buffer-rope and buffer management within your organization where it can do the most good. The case studies provided in this handbook can be used as a model.

THE FUTURE

In Section III of this handbook, Chapter 10 describes the performance measurement system that should replace the problem-prone traditional system. Chapter 11 describes a new problem-solving approach called the Thinking Processes. This last section provides a solid base from which implementation and continued success using constraints management can be achieved.

This questionnaire is to provide information on your *plant's* operation and its inventory/ production management system.

1. Your industry is _____

2. Your primary product line is _____

3a. Your plant is independent   ®
    A division of a corporate   ®
    Part of a corporate division   ®   Other (specify) _____

3b. Does your plant have profit and loss responsibility?   Yes   ®   No   ®
    Does your plant have marketing responsibility?   Yes   ®   No   ®
    Does your plant have pricing responsibility?   Yes   ®   No   ®

4. The approximate annual dollar sales volume of this plant is _____

5. The total number of plant employees is _____
    Of these, how many are salaried?   _____
    Of these, how many are hourly?   _____
    Of these, how many are direct?   _____
    Of these, how many are indirect?   _____

6. The following three items ask for a description of various characteristics of your operation. Please estimate how your operation is best portrayed. (Indicate the percent on each line; your responses for each item should sum to 100%.)

    a. Customer Base (sum to 100%)
      _____% make-to-stock
      _____% make-to-order
      _____% engineer-to-order
      _____% assemble-to-order
      _____% other (please describe)_____

    b. Volume (sum to 100%)
      _____% each item is engineered/designed for the customer
      _____% intermittent batch (low volume, many models)
      _____% intermittent batch (high volume, fewer models)
      _____% continuous (high volume flow)
      _____% other (please describe)_____

Figure 6.1   Participant's Previsit Questionnaire

c. Manufacturing Type (sum to 100%)
_____% discrete (individual units)
_____% process (volume reporting)
_____% other (please describe)_____
_____

7. What is the *manufacturing* lead time for a typical customer order? _____

8. Number of different finished products manufactured_____

9. Number of different end items in the master schedule _____

10. Number of different component items (including subassemblies, parts, and raw materials) in the inventory system_____

11. Number of levels in the typical end item bill of material in your primary product line _____

12. Number of different work centers (including assembly stations) in your plant __

13. Standard operating policy of plant operation is:
(check one)   ® 5 days per week   ® 6 days   ® 7 days
(check one)   ® 1 shift   ® 2 shifts   ® 3 shifts   ® 4 shifts

14. What is the fraction of your total inventory investment in each of the following categories? (Indicate the percent on the line; should sum to 100%.)
_____% raw materials   _____% work-in-process   _____% finished
_____% other (please specify)_____

This portion is concerned with your production/inventory control system(s). Mark all the elements that your plant routinely uses. It is likely that more than one production/inventory control system is used in your operation. Also indicate those parts that do not apply to your operation. Indicate beside each element if it is currently used (Used), being implemented now (Impl), or not applicable (N/A).

		Used	Impl	N/A
1.	Material Requirements Planning System (MRP)	®	®	®
a.	Aggregate Capacity Planning	®	®	®
b.	Master Production Schedule (MPS)	®	®	®
c.	Material Requirements Planning (MRP)	®	®	®
d.	Capacity Requirements Planning (CRP)	®	®	®
e.	Shop Floor Control System (SFC)	®	®	®
f.	Purchasing Scheduling System	®	®	®

Please indicate:
MRP implementation status: Date started _____
Fully operational   ®   Implementation in process   ®

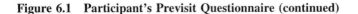

Figure 6.1   Participant's Previsit Questionnaire (continued)

	Used	Impl	N/A
2. Manufacturing Resource Planning System (MRPII)	®	®	®
a. Marketing uses MRP promise dates	®	®	®
b. Accounting uses MRP data for budgets	®	®	®
c. Engineering uses MRP data for changes	®	®	®
d. Vendor Scheduling	®	®	®
e. Distribution Requirements Planning (DRP)	®	®	®
f. Other (explain) _____			

Please indicate:

MRPII implementation status: Date started _____

Fully operational ®   Implementation in process ®

	Used	Impl	N/A
3. Theory of Constraints	®	®	®
a. TOC Software	®	®	®
b. Principles, T, I, and OE used	®	®	®
c. Drum, Buffer, Rope (DBR) used for SFC	®	®	®
d. V-A-T Analysis	®	®	®
e. Theory of Constraints Five-Step Process	®	®	®
f. Effect-Cause-Effect and Evaporating Clouds	®	®	®
g. Disaster Software	®	®	®

Please indicate:

TOC implementation status: Date started _____

Fully operational ®   Implementation in process ®

	Used	Impl	N/A
4. Just-in-Time System (JIT)	®	®	®
a. Level Master Schedule	®	®	®
b. Pull Method of Production—Kanban	®	®	®
c. Focused Factory within a Factory	®	®	®
d. Revised Plant Layout	®	®	®
e. Reduced Lot Sizes	®	®	®
f. Setup Time Reduction Program	®	®	®
g. Multiskilled Workers	®	®	®
h. Schedule Linearity	®	®	®
i. Quality Emphasis Program (TQC)	®	®	®
j. Scheduled Preventive Maintenance	®	®	®
k. Supplier Capacity Control	®	®	®
l. Supplier Involvement with New Product Design	®	®	®
m. Other (explain) _____			

Figure 6.1 Participant's Previsit Questionnaire (continued)

Please indicate:

JIT implementation status: Date started _____

Fully operational ®   Implementation in process ®

5. The most critical resource to meet schedules in generally (please rank 1–5)

_____ materials availability   _____ capacity availability

_____ energy availability   _____ manpower availability

_____ other (please specify) _____

6. Briefly describe your production process: _____

_____

_____

_____

7. For each item in the following list of plant and environmental characteristics, please check if this characteristic is present in your operation. If it is, indicate the importance of the characteristic by circling the appropriate number.

0	1	2	3	4	5
Not Applicable	Unimportant		Somewhat		Extremely Important

Capacity Flexibility	0	1	2	3	4	5
Complexity of Product and Process Flow Design	0	1	2	3	4	5
Length of Lead Time for Increasing Capacity	0	1	2	3	4	5
Difficulty in Scheduling Preventive Maintenance	0	1	2	3	4	5
Deep (Multilevel) Bill of Material	0	1	2	3	4	5
Long Production Runs	0	1	2	3	4	5
Lead Time for Acquiring Raw Materials	0	1	2	3	4	5
Shelf Life of Raw Materials	0	1	2	3	4	5
Compliance with Government Regulations	0	1	2	3	4	5
High Volume of Production	0	1	2	3	4	5
Unique Routings for Components	0	1	2	3	4	5
Storage Capacity Limitations	0	1	2	3	4	5
Seasonal Demand Patterns	0	1	2	3	4	5
High Setup Costs Associated with Switching Products	0	1	2	3	4	5
Need for Lot Traceability	0	1	2	3	4	5
Low Number of End Items	0	1	2	3	4	5
Large Number of Products Having the Same Routings	0	1	2	3	4	5
Large Number of Assemblies Having the Same Routing	0	1	2	3	4	5
Job Shop (Functional) Layout	0	1	2	3	4	5

Others (please specify characteristics and importance) _____

_____

_____

Figure 6.1   Participant's Previsit Questionnaire (continued)

8. The following functions have been identified by researchers as possible parts of a planning and control system. Please check the appropriate blanks to indicate those that are components of your production planning and control system and indicate whether it is a manual or computerized part of the planning and control system.

	Not Used	Manual	Computerized
Forecasting	®	®	®
Material Requirements	®	®	®
Aggregate Planning	®	®	®
Master Production Scheduling	®	®	®
Rough-Cut Capacity Planning	®	®	®
Detailed Capacity Planning	®	®	®
At Key Work Centers	®	®	®
Material Release	®	®	®
Dispatch List Scheduling	®	®	®
Shop Floor Control	®	®	®
Capacity Control	®	®	®
Input/Output Control	®	®	®

Figure 6.1 Participant's Previsit Questionnaire (continued)

Part A: Background Information from the Previsit Survey

Firm Name: _____   Product(s): _____

Production Planning and Control System Used: _____

Name and Title of Interviewee: _____

Part B: Description of Production Planning and Control System

1. Please describe the physical layout of the facility under study.
    a. What part is a functional layout?
    b. What part is a product layout?
    c. Are any parts cellular?
    d. Describe any combination.

2. How was the layout developed?

3. How long has this layout been in place?

4. Describe the actual manufacturing department or work center under study.
    a. Number of work centers?
    b. How are the work centers related? (Routing)

Figure 6.2 On-Site Interview

c. How is the department supervised?
d. What is the reporting relationship?

5. Describe the process as components for the end product are manufactured.
   a. Are the number of end items relatively large compared to raw materials?
   b. Are all end items produced in essentially the same way?
   c. Are all components produced using dedicated machine tools?
   d. Are components unique to specific end items?
   e. Are routings for components unique or highly dissimilar?
   f. Are there points in the routings where components are physically committed to an end item? (Point of divergence)
   g. Is there typically a large amount of inventory in front of any bottleneck operation? (Look for misallocation)
   h. Are there a large number of subassemblies going into a few end items?
   i. Are components unique to specific end items?

6. Describe the interaction between this department and other departments?
   a. Quality assurance
   b. Quality control
   c. Materials management
   d. Purchasing
   e. Physical engineering
   f. Manufacturing engineering
   g. Design engineering
   h. Accounting
   i. Marketing and sales

7. Describe the software package used for the production planning and control system:
   a. Purchased or developed in-house?
   b. If purchased, name of the supplier?
   c. How was the decision made to make or buy?
   d. How satisfied are you with the validity of the output?
   e. How useful is the output in making work decisions?
   f. How timely is the output from the system?

8. Describe the educational programs used in the implementation of the system:
   a. Upper management
   b. Manufacturing
   c. Materials
   d. Other functional areas
   e. Customers
   f. Line employees
   g. When did the educational program occur?

Figure 6.2　On-Site Interview (continued)

    h. Was the program on compensated time
    i. Is the educational program ongoing

9. Estimate the accuracy of a typical bill of material.
    a. How often are the bills changed?
    b. What is the process when the bills are changed?
    c. What problems exist with the accuracy and validity of the bills?

10. Estimate the accuracy of the part routings.
    a. How often are the routings changed?
    b. What are the primary reasons for the changes?
    c. What is the process when the routing is changed?
    d. What problems exist with the accuracy and validity of the routings?

11. Describe the inventory management practices used.
    a. Has an ABC analysis been performed? What criteria were used?
    b. Is cycle counting used to manage the inventory accuracy?
    c. If a physical inventory is used, how is it conducted?
    d. How are the inventory items identified on the shop order?
    e. How is safety stock used in the department?
    f. What kind of order policies (lot sizes) are used?
    g. How are the order polices developed?
    h. Describe any problems with inventory management or policies.
    i. Describe the performance systems used to manage the inventory.

12. Describe the factory's relationship with the customers.
    a. Who are the chief customers for the products under study?
    b. How is contact made to the customers from the factory?
    c. How do the customers contact the factory (EDI Or 800 numbers)?
    d. What are the customers' chief questions (due dates, status)?
    e. What problems occur with customer contacts?

13. Describe how the performance of the following functions is measured.
    a. The overall factory
    b. The plant manager
    c. The master production schedule
    d. The materials function
    e. The manufacturing function
    f. Supervisors
    g. Line employees
    h. Transportation
    i. Purchasing
    j. Vendors

Figure 6.2   On-Site Interview (continued)

Part C: Master Production Scheduling

1. How is the forecast used in creating the MPS?

2. Describe how a customer order is processed.

3. Describe the master scheduling system being used.

4. What items are included in the MPS? If not all, then what criteria are used?

5. How frequently is the MPS updated?

6. What is the time bucket used? How was this selected among alternatives?

7. How is the MPS linked to the production plan?

8. What units are specified in the production plan?

9. Are any mathematical modeling techniques used, such as linear programming, to develop the schedule? If so, how are they used?

10. Which components of the MPS are done manually?

11. Which components are done by computer?

12. How is any MPS backlog managed? What is the average backlog?

13. What is the approval process for the MPS?

14. What are the problems associated with developing the MPS?

15. Are "what if" questions answered by simulations?

16. What logical structure characteristics were important in determining the procedures used for developing the MPS? What impact have these had on the procedures?

17. What physical layout and process characteristics were important in determining the procedure used for developing the MPS? What impact have these had on the procedures?

18. What policy characteristics were important in determining the procedure used for developing the MPS? What impact have these had on the current procedures?

19. What supporting MIS characteristics were important in determining the procedure used for developing the MPS? What impact have these had on the current procedures?

20. Describe what happens to the MPS when there is a short lead change in requirements.

Priority Planning

1. How is Material Requirements Planning (MRP) (or Just-in-Time or TOC) used to determine vendor and shop order release dates for component times?

Figure 6.2 On-Site Interview (continued)

2. If MRP (JIT, TOC) is not used, how are material requirements determined for component items?

3. What components of this process are done by computer?

4. What components of this process are done manually? Why?

5. What information is required to determine these requirements?

6. How frequently does this take place?

7. What are problems associated with determining these requirements? How are they dealt with?

8. What logical structure characteristics were important in determining the procedures used for keeping priorities valid? What impact have these had on the current procedures?

9. What plant layout and process characteristics were important in determining the procedures used for keeping priorities valid? What impact have these had on the current procedures?

10. What policy characteristics were important in determining the procedures used for keeping priorities valid? What impact have these had on the current procedures?

11. What supporting MIS characteristics were important in determining the procedures used for keeping priorities valid? What impact have these had on the current procedures?

Capacity Planning

1. How is capacity defined and measured?

2. What type of long-term capacity planning procedure is used?

3. Are capacity bills or resource bills used?

4. What level of aggregation is used in long-term capacity planning?

5. What time horizon is used in long-range capacity planning?

6. How frequently is long-range capacity planning done?

7. What types of decisions are made from the long-range capacity plan?

8. When annual requirements are expected to exceed capacity, what adjustments are made to increase output?

9. When annual requirements are expected to be less than capacity, what adjustments are made to reduce output?

10. In the medium and short term, how is the master production schedule checked against capacity?

Figure 6.2   On-Site Interview (continued)

11. How frequently is this done?

12. What is this information used for?

13. Which pieces of equipment are checked?

14. What are the information inputs to this process?

15. What happens when short-term requirements are seen to exceed capacity?

16. What happens when short-term requirements are seen to be less than capacity?

17. What problems are associated with capacity planning? How are they dealt with?

18. What logical structure characteristics were important in determining the procedure used for capacity planning? What impact have these characteristics had on the current procedures?

19. What physical layout and process characteristics were important in determining the procedure used for capacity planning? What impact have these characteristics had on the current procedures?

20. What policy characteristics were important in determining the procedures used for capacity planning? What impact have these characteristics had on the current procedures?

21. What supporting MIS characteristics were important in determining the procedures used for capacity planning? What impact have these characteristics had on the current procedures?

Priority Control

1. Once orders are released to the shop floor, how are order priorities adjusted to keep up with MPS changes?

2. How does the final assembly schedule relate to the MPS?

3. How are adjustments made to the plan?

4. Is a daily dispatch list part of your planning and control system? How is this accomplished?

5. When several orders are open, how are relative priorities determined?

6. Is the backlog at the bottleneck work center controlled by a daily dispatch list? How does it consider available work center capacity? Who decides?

7. What techniques are used to manage queues at the bottleneck?

8. What parts are typically on the hot list?

9. What are common problems associated with keeping priorities valid? How are they dealt with?

Figure 6.2   On-Site Interview (continued)

10. What logical structure characteristics were important in determining the procedures used to keep priorities valid? What impact have these characteristics had on the current procedures?

11. What layout and process characteristics were important in determining the procedures used to keep priorities valid? What impact have these characteristics had on the current procedures?

12. What policy characteristics were important in determining the procedure used to keep priorities valid? What impact have these characteristics had on the current procedures?

13. What supporting MIS characteristics were important in determining the procedure used to keep priorities valid? What impact have these characteristics had on the current procedures?

Capacity Control

1. In which work centers is capacity monitored?

2. How is equipment utilization monitored and measured?

3. How frequently are these measurements monitored and measured?

4. How is manpower utilization monitored and measured?

5. What procedures are used to schedule overtime?

6. How frequently are these measurements made?

7. How frequently are they reported?

8. What are common problems associated with capacity control? What impact have these characteristics had on the current procedures?

9. What logical structure characteristics were important in determining the procedure used to control capacity? What impact have these characteristics had on the current procedures?

10. What layout and process characteristics were important in determining the procedure used to control capacity? What impact have these characteristics had on the current procedures?

11. What policy characteristics were important in determining the procedure used to control capacity? What impact have these characteristics had on the current procedures?

12. What supporting MIS characteristics were important in determining the procedures used to control capacity? What impact have these characteristics had on the current procedures?

Figure 6.2   On-Site Interview (continued)

REFERENCES

1. Marshall, C. and G.B. Rossman, *Designing Qualitative Research*, Newbury Park, CA: Sage Publications, Inc., 1990.
2. Patton, M.Q., *Qualitative Evaluation and Research Methods,* 2nd ed., Newbury Park, CA: Sage Publications, Inc., 1990.
3. Babbie, E., *The Practice of Social Research,* 5th ed., Belmont, CA: Wadsworth Publishing Co., 1989.
4. Meredith, J.R., A. Raturi, K. Amoako-Gyampah, and B. Kaplan, "Alternative Research Paradigms in Operations," *Journal of Operations Management,* 8(4), 297-326, 1989.
5. Flynn, B.B., S. Sakakibara, R.G. Schroeder, K.A. Bates, and E.J. Flynn, "Empirical Research Methods in Operations Management," *Journal of Operations Management,* 9(2), 250-284, 1990.
6. Eisenhardt, K.M., "Building Theories From Case Study Research," *Academy of Management Review,* 14(4), 532-550, 1989.
7. Strauss, A.L., *Qualitative Analysis for Social Sciences,* Cambridge: Cambridge University Press, 1990.
8. Huck, S.W., W.H. Cormier, and W.G. Bounds, *Reading Statistics and Research,* New York: Harper & Roe, 1974.

VALMONT/ALS,* BRENHAM, TEXAS

This chapter illustrates how a wholly owned corporation within a larger corporation was able improve its competitiveness in relation to similar facilities and by so doing become a preferred manufacturing site within the parent corporation.

BACKGROUND

Valmont/ALS, located in Brenham, Texas, is a manufacturer of lighting poles and electrical support structures. The plant in Brenham is one of five plants that comprise the Industrial and Construction Products Division of Valmont Industries, Inc. The parent company, Valmont Industries, Inc., has its headquarters in Valley, Nebraska. Prior to being acquired in 1980 by Valmont Industries (the plant in Brenham was an independent company), American Lighting Standards, hence the initials ALS in the plant's name. Two other Valmont plant locations, Tulsa, Oklahoma and Valley, Nebraska, also fabricate light poles and have similar production capabilities.

The product line at the Brenham plant is classified into two categories: small poles and large poles. Small poles are manufactured from purchased, tapered round, or square steel tubing up to an equivalent diameter of 12 inches. The small poles are used to support light fixtures along interstate highways, in parking lots, along city streets, and other similar applications.

---

* This case was originally developed by James L. Wahlers[1] and was modified for this text with his permission.

The large poles are engineered and fabricated to order from steel plate in thicknesses up to five-eighths of an inch. The large poles are used as supports for stadium lighting, electrical transmission towers, traffic light support systems, and other comparable applications. Some large pole applications require a completed fabrication well over 100 feet in height. These poles are fabricated in two or more sections and are designed so that the sections telescope into one another. Both the small pole and large pole product lines are engineer-to-order. Many transmission applications require separate engineering on each individual pole due to variations in terrain and changes in direction of the transmission line.

Sales from the Brenham plant are estimated to have increased nearly 65% in the last two years. Seventy hourly and 20 salaried workers are employed at this location. The plant has 40 workstations that are classified as individual cost centers by the traditional accounting system definition.

Plant Operations

The Brenham plant is divided into two departments. One department is concerned with the fabrication of the small poles and the other with the manufacture of the large poles. The small pole fabrication department manufactures the accessories that are attached to the pole and become part of the pole assembly. These accessories include base plates, fixture support arms, wiring access panels, etc. The large pole department requires the same accessories and component parts; however, the large tubes themselves are fabricated in the shop. Pieces of steel plate are flame or plasma cut to the proper flat shape. A computer-aided design (CAD system) inputs the flat shape into a numerically controlled combination flame/plasma cutting machine. The large poles are then formed from the flat, cut shape into a tapered cylinder in a large press brake. The smaller sizes of the large poles, under 30 inches in diameter, are formed into a tapered cylinder and automatically seam welded with a single longitudinal seam. Larger diameter poles, from 30 inches to a maximum of 57 inches in diameter, are formed into two tapered half cylinders. The halves are also automatically welded together but with two longitudinal seams. Seam welds are 100% ultrasonically tested for welding or other defects. All seam welding defects are immediately repaired manually by certified welders. The seam welding operation is the major source of defects and the cause of subsequent rework in pole manufacturing.

The accessories and components required for both the large poles and small poles are fabricated through the same work centers. The components required for small poles and large poles vary in size and weight but belong generally to the same parts family for similar applications on the finished poles.

The component parts proceed to the respective assembly areas for the final assembly process. The small poles and the large poles are assembled in separate assembly areas. The assembly area has been determined to be the constraint resource in each of the product lines. After assembly, the poles proceed to the final finishing area. In the final finishing area, all critical welds are ultrasonically tested and the assembled poles are inspected. Any repairs are performed in the finishing area prior to the next process. The customers have three options for the finish on their poles. Poles can be shipped with a natural oxidized finish that is inherent in a special steel plate, called "Corten" steel, specified when the poles are ordered. In addition, the poles may be painted or galvanized. Painting is performed at the Valmont site; the galvanized finish is provided by outside contractors. Poles are shipped to the galvanizers and are shipped from the galvanizers directly to customers. All other shipments to customers are made from the Brenham site.

The hourly work force is nonunion and made up primarily of machinists and welders. Because of the critical nature of the welding operation in the pole assembly and fabrication process, welding is considered the critical skill in the Brenham plant. Because certified welders are in short supply, Velmont conducts an extensive training program to develop and train its welders. The plant operates two, or more exactly, one-and-one-half, shifts five days a week. The main production is performed on the day shift. The second shift provides time to build any components required to fill voids or holes in the constraint buffers.

THEORY OF CONSTRAINTS IMPLEMENTATION

Valmont/ALS has a fully functional MRP system. It was not known exactly when the system had been installed, but the production control manager's best estimate was that the system had been installed 10 or 15 years prior to the time of this interview. Using the MRP system, the plant was scheduled in a traditional manner. Production batches were large; emphasis was on labor efficiency and machine utilization. Orders routinely were overrun to improve efficiency and avoid frequent machine setups. The rationale was that any parts that were produced would be used some time in the near future. No records were kept that verified when these components were in fact used. Work-in-process inventory was unacceptably high; expediting was the rule. The plant made long runs to improve efficiency, but did not necessarily work on the parts that were needed to fill the immediate orders. Orders were frequently late, and because of the inability to accurately predict delivery dates, orders were regularly switched among the Valley plant, the Tulsa plant, and the Brenham plant. Consequently, the level of work in the plant varied greatly; layoffs were common.

In the fall of 1986, the plant general manager and his staff, after reading Goldratt's book *The Goal*,[2] became enthusiastic about its concepts. This enthusiasm led them to purchase several copies of Goldratt's second book, *The Race*[3] (written with Bob Fox). In addition to reading *The Race,* several staff members attended seminars at the Goldratt Institute. A simulation game, called The OPT Challenge, which runs on a personal computer, was also purchased from the Goldratt Institute. An understanding of the principles of optimized production technology (OPT) and the further refinement called the Theory of Constraints was developed both from attending the seminars at the Goldratt Institute and running the PC simulations. Late in the first quarter of 1987, the management of the Brenham plant decided to implement some of Goldratt's principles.

The initial step taken was to reduce batch sizes approximately one-half at the gating resources of the small pole operation. A "dramatic improvement in the flow of material through the shop" was reported when this change took place. No measurements were made or quantitative records kept, but less expediting and improved due date performance were reported by staff members.

Encouraged by the results of the first trials, the management of the Brenham plant decided to implement more aspects of the Theory of Constraints principles. In addition, training efforts were expanded to the entire hourly and salaried complement. The Brenham location became the "Beta site" to determine the applicability of Goldratt's theories to this division of Valmont Industries. The plant general manager, along with key staff members, attended various seminars and training sessions at the Avraham Y. Goldratt Institute. After the staff completed this training, the entire salaried complement was given training sessions developed by the Goldratt Institute that were appropriate for the individual disciplines within the operation. The two-day training sessions relied heavily on the use of personal computers. The plant management was concerned that the hourly workers would be intimidated by the computers. To train this portion of the work force, a special program using a dice game, which is featured in both *The Goal*[2] and *The Race,*[3] was developed and used to demonstrate how material flows through a factory.

While the training was proceeding, operational changes were under way at Brenham. The assembly areas for both the small and the large poles were identified as the constraints. The production control manager began to manually forward schedule the constraint and then use the existing MRP system to backward schedule the unconstrained work centers. At the same time, the plant instituted a policy of only building component parts for poles to fill a customer order. The policy of building components based upon a forecast was eliminated. This change in policy required that reassurances be given to the hourly work force to allay concerns that the elimination of work-in-process inventory was by design and not because of a lack of future business.

The plant staff gradually instituted additional segments of the Theory of Constraints as they became comfortable with the portions that had previously been implemented. Gradual implementation continued through 1987.

In early 1988, the plant began to study the impact of buffers on the performance of the constraint. The plant has an accurate database containing routings and standard hours. This database is a part of the existing MRP system. A proprietary software system called Data Query made it possible to access this database and accurately and quickly track the status of individual parts required at the constraint to ensure the timely completion of an order. The plant established a six-day time buffer ahead of the constraint, the pole assembly area. Region 1 of the buffer contained the two days' work immediately prior to the constraint. This portion of the buffer was the most critical because the components represented by those two days were the material required to complete the jobs being processed by the constraint. The material that would be required for days 3 and 4 represented region 2 of the buffer. Components of lowest immediate priority comprised region 3 and represented days 5 and 6.

Using the data query system, it was possible to determine which material was not available in the buffer when it was required. The highest priorities were established for material required in region 1. Analysis provided information specifying the work centers that were the most critical in maintaining the constraint buffers. A system of measuring the performance of the constraint and the critical work centers was developed. This one-page report is available by 9:30 each morning (see Figure 7.1). The purpose of the report

SUBJ.: SHOP PERFORMANCE REPORT FOR WEEK ENDING: 15-JULY   AS OF : 07/17   09:24 AM

CONSTRAINT PERFORMANCE

CC # 0139 SMALL POLE ASSEMBLY

	MON.	TUES.	WED.	THURS.	FRI.	SAT.	WK TO DATE	MON. TO DATE
SCHEDULED STD. HOURS	63.00	68.00	68.00	68.00	65.00	0.00	340.00	556.00
STD. HOURS RECOVERED	44.99	26.75	47.50	47.52	44.50	0.00	211.26	368.02
ACTUAL HOURS WORKED	48.25	68.68	97.98	96.58	78.25	0.00	389.75	552.07
PROD. % (STD HRS / HRS WORKED)	93.24%	38.95%	48.48%	49.20%	56.87%	0.00%	54.20%	66.66%
SCHED. % (STD HRS / SCHED HRS)	65.12%	39.34%	69.85%	69.88%	65.44%	0.00%	62.14%	66.19%

CC #1495 LARGE POLE FIT & # 242 ARM LAYOUT & TACK WELD—TP

	MON.	TUES.	WED.	THURS.	FRI.	SAT.	WK TO DATE	MON. TO DATE
SCHEDULED STD. HOURS	30.00	30.00	30.00	30.00	30.00	0.00	150.00	206.00
STD. HOURS RECOVERED	27.12	17.40	30.54	18.11	29.66	0.00	122.83	166.38
ACTUAL HOURS WORKED	32.00	26.08	29.84	19.75	24.00	0.00	131.67	178.67
PROD. % (STD HRS/HRS WORKED)	84.75%	66.72%	102.35%	91.70%	123.58%	0.00%	93.29%	93.12%
SCHED. % (STD HRS/SCHED HRS)	90.40%	58.00%	101.80%	60.37%	98.87%	0.00%	81.89%	80.77%

REGION 1 BUFFER ANALYSIS

TODAY'S BUFFER HOLES, POTENTIAL LOST THRUPUT; W/O SET-UP

COST CENTER RESPONSIBLE	MON. 9:30AM	TUES. 9:30AM	WED. 9:00AM	THURS. 9:30AM	FRI. 9:30AM		WK TO DATE	MON. TO DATE
0140 UPRIGHT DRILL PRESS	0.06	0.46	0.00	0.00	0.00		0.52	1.12
0143 RADIAL DRILL PRESS	0.00	0.19	0.00	1.16	3.00		4.35	11.12
0643 PARTS FIT-UP & WELD	0.00	0.00	0.00	0.00	0.00		0.00	10.37
0778 TANDEM PRESS BREAK	0.00	0.00	0.39	0.00	0.00		0.39	2.50
0784 PLASMA CUT-A-LINE	0.00	0.00	0.00	0.00	3.11		3.11	3.11
0789 SEAM-WELD	0.00	0.00	0.75	0.00	6.17		6.92	18.88
0928 PANTOGRAPH	0.02	0.26	0.00	0.18	0.06		0.52	10.37
0929 DESLAG, DEBURR, GRIND	0.02	0.16	0.00	0.13	0.45		0.75	12.37
1497 SEAM REPAIR	0.00	0.00	0.41	0.00	6.46		6.87	29.37
OTHER MISC. COST CENTERS	0.00	0.00	0.20	0.06	5.01		5.27	18.73
TOT. MISS. HRS. R1 BUFFER	0.10	1.07	1.75	1.53	24.26		28.69	117.93

REGION 2 BUFFER ANALYSIS

	MON.	TUES.	WED.	THURS.	FRI.		WK TO DATE	MON. TO DATE
TOT. HRS REQ. FOR R-2 BUFFER	124.61	134.39	145.28	119.01	73.06	AVE:	119.27	251.37
HRS.COMPLTD IN R-2 BUFFER	62.62	76.52	82.08	48.20	38.52	AVE:	61.58	117.89
% OF COMPLTD OF R-2 BUFFER	50.2%	56.9%	56.5%	40.5%	52.7%	AVE:	51.6%	46.9%
STD HRS. RECOV. DEPT 2151	48.71	68.12	72.22	66.30	109.28	0.00	364.63	663.27

THROUGHPUT PERFORMANCE

THRUPUT (N. SALES- MAT'L)	$53,525	$3,948	$68,951	$66,224	$0		$192,648	$232,265

Figure 7.1   Constraint Performance Report

is to provide the managers immediate information on buffer shortages by work center. Shop supervisors and personnel utilize an additional report, available by 7:30 A.M., that provides detailed information by part number. Shortages or "holes" in the region 1 buffer require immediate corrective action to avoid interruption of the constraint operation.

The region 2 buffer analysis is not broken down by individual work center, but rather only by the percentage of work completed. The control figure for region 2 is 50%; that is, if 50% of the work in region 2 is complete, the buffer is in satisfactory condition. This method of analysis is consistent with the fact that only the components in region 1 are critical to the completion of an assembly and that is where the detailed control is required.

Region 3 of the buffer is not measured. The measurements in region 1 and region 2 provide sufficient information for adequate control.

A shipping buffer of two weeks has been established. Orders are targeted to be completed two weeks prior to the customer due date. This provides the opportunity to ship orders ahead of schedule if the customer can accept the material or on occasion allows the firm to accept an order with short lead time requirements, adjust the production operation, and still ship existing orders on time.

PRODUCTION PLANNING AND CONTROL SYSTEM

Sales of light poles are handled by four regional sales managers; however, Valmont does not have a regular sales force but rather is represented in the field by manufacturer's representatives. Valmont's customers are primarily distributors of electrical or lighting products and original equipment lighting manufacturers. Customer orders are sent to the headquarters and entered into the Open Order Sales Center. From this point, the order is sent to the engineering department, where the detailed engineering is performed and the bills of material for the individual poles is developed. Poles are engineered individually; however, a project is under way to develop preengineered categories of poles to expedite the engineering process. After engineering is completed the orders are assigned to the manufacturing locations based upon both geographic considerations and plant workload. Standard lead times are quoted to the customers. The manufacturing lead times begin after engineering is completed. At the time of this interview, the Brenham plant was quoting a lead time of four weeks for the small pole operation and six weeks for the large poles. If there is a requirement from a customer for a delivery date that is sooner than the standard lead time, the headquarters sales department inquires if the plant can ship earlier. If the plant can meet the required date, the order is sent to the plant for immediate fabrication.

When an order is received in Brenham, the basic MRP database is utilized to net any requirements against the on-hand inventory balances of all components required for that individual order. The shipping date is established by forward scheduling the order from the constraint. If the customer will accept the order, it is shipped based upon the plant's completion schedule. If the order cannot be shipped early, the order completion is generally scheduled two weeks prior to the required date. After the constraint schedule is determined, the work centers that support the order are backward scheduled from the completion date established by the constraint schedule. The data query system is used to monitor the progress of an order on a daily basis to assure that the buffers are being maintained. Since the buffers total approximately one-week's time, the plant is actively working on only material required in the next week. When holes appear in the buffers, resources are immediately diverted to the correct parts required to fill those holes.

The MRP system is not used for day-to-day shop floor scheduling; it is used to plan the purchase of long lead time and routine component items. The tapered round and square poles used in the small pole assembly are purchased and have a lead time of six to eight weeks. Similarly, the steel plate required for the large pole fabrication has a relatively long lead time. These materials, along with purchased hardware that is part of the assemblies, are purchased based upon the MRP system.

The databases maintained by the Brenham plant are very accurate. The accuracy of the bills of material is between 95 and 98%. Similarly, the accuracy of the inventory records is within 1% of the headquarters auditors. The accuracy of these databases provides the basis for schedule accuracy.

Detailed records of shipment to schedule or inventory reduction have been maintained by the plant. The production control manager states that due date performance exceeds 93%. Previously, due date performance averaged in the mid-80% range. Work-in-process inventory is now maintained at six days, the content of the buffers. Raw material inventory has been increased slightly to enhance the plant's ability to provide short lead times on special customer requirements. Overall inventory levels are nearly the same as in the three previous years; however, this level of inventory now supports a sales/shipping volume that has increased 65%. Additionally, the mix of material held in inventory has changed. Inventory now consists of primarily raw material and purchased components. Work-in-process has been reduced to the size of the constraint buffers. Finished goods inventory is limited to the contents of the shipping buffer.

COST ACCOUNTING SYSTEM

Prior to the implementation of the Theory of Constraints, the plant used a traditional standard cost accounting system to develop the cost of its products. Johnson and Kaplan[4] suggest that management accounting systems must perform three different functions: (1) meet the firm's external financial reporting requirements, (2) provide the relevant information needed to plan and control the facility, and (3) provide the intelligence needed to realistically price a firm's products. The performance of these three functions may dictate the use of multiple accounting systems. The strategy suggested by Johnson and Kaplan[4] is the basis for the method the Brenham plant has used. The established, traditional job order costing system has been maintained with minor modifications and is employed for the financial reporting required by the corporation.

However, significant changes have been made in how the costs are determined by the plant for its internal decision making and profitability measurement. For reporting inventory valuation and standard costs, the traditional system is used. By considering nearly all plant costs as fixed costs, it is possible to set the burden rates in such a manner that standard costs are reported with negligible variances. The goal of the plant's system is to eliminate positive or negative variations wherever possible.

The method used to determine the profitability of the manufactured products is also based upon the assumption that all plant labor and overhead costs are fixed over the time period being considered. The validity of this assumption is supported by empirical data developed by the plant controller over a two-year period. The amount of time a product is processed at the constraint is the main criterion in determining the profitability of the product. The cost per constraint hour is determined by dividing the total plant cost (direct labor plus all other plant costs) by the number of hours the constraint is operated. Throughput is defined as net sales (total sales minus commissions, transportation, and materials). The same reasoning is used when a cost estimate for a product is prepared. The profit rate for that product is calculated by dividing the net throughput by the constraint time required. The estimated product cost is based upon the standard hours an order will spend at the constraint.

Decisions affecting proposed investments to increase capacity or reduce cost are based upon the effect the expenditure will have upon the plant throughput. An increase in the capacity of the constraint or the nonconstraint's protective capacity is critical to the decision-making process. Protective capacity is defined as the capacity of the individual work centers that are found to be critical in maintaining the constraint buffer. The support resources at Valmont/ALS are the work centers in the region 1 buffer analysis portion of the report. Projects that do not impact these work centers or the constraint itself are not considered desirable expansion or cost reduction projects by the management of the Brenham plant, even though traditional analysis would indicate the expenditure was justified.

PERFORMANCE MEASUREMENT

Prior to the implementation of the Theory of Constraints, the Brenham plant of Valmont Industries used traditional performance measures. Labor efficiency and equipment utilization were of paramount importance in manufacturing and were monitored closely by plant management. Supervisors routinely were concerned with their own or their department's efficiency or utilization performance rather than the performance of the division. Stressing these measures resulted in excessive quantities of work-in-process inventories. Expediting was common; due date performance was less than 85%. Because of the inability of the plant to satisfy customer requirements, the plant workload varied and layoffs were common.

This plant is actually a separate corporation that is wholly owned by the parent corporation, Valmont Industries, Inc. The traditional externally reported measures of corporate performance (profit, return on investment, and return on owner's equity) are routinely reported to the corporation's headquarters in Valley, Nebraska. Profits are calculated as previously described. Both return on investment and return on owner's equity are calculated in the traditional manner.

Internal operating performance is measured daily using the information contained on the one-page report. The management of the Brenham plant considers that this one-page daily report contains the information required to plan, measure, and control performance.

The constraint performance measure is provided at the top of the report. The scheduled standard hours, the standard hours recovered, and the actual hours worked for the small pole and the large pole assembly areas are displayed. The goal is to keep the production and scheduled percentages as near to 100% as possible.

The performance of the constraint buffers is provided next. The plant personnel have determined that nine cost centers are critical to maintaining the operation of both the small and large pole constraints. These work centers are listed in the region 1 portion of the report. This information is current for that day and is available on the shop floor in more detailed formats by 7:30 each morning. A number other than zero at a cost center designates a "hole" in the buffer and requires immediate corrective action by the buffer supervisor. If more than 20 hours is missing from the region 1 buffer, a serious problem exists. Twenty hours is approximately one-third of the total buffer time. Since it is difficult to make this time up, it is likely that an order will be late.

The region 2 buffer is analyzed in much less detail on this report. The completed components represented by the hours in this buffer will not be required at the constraint for three to four days. The cost control centers are not listed. The operating goal is to have between 50 and 60% of hours for the region 2 buffer complete each day. The plant people have determined that this level of completion provides adequate protection for the region 1 buffer. The region 3 buffer, representing material not needed for five to six days, is not listed on this report. These are the components that are currently being released to the gateway work centers. The flows are continuously monitored to determine if there are other work centers in the system that should be considered "protective capacity," defined as resources that are most nearly fully utilized or are the source of "holes in the region 1 buffer."

The final portion of this report summarizes daily throughput performance. Throughput is defined as sales minus the sum of material, transportation, and commissions (in economics, true variable costs). The figure on the report reflects the daily product shipments to customers. The week-to-date and month-to-date figures reflect the weekly and monthly totals. Throughput performance is not used as a control measure. It is a "scorekeeping" measure that informs management of progress toward covering the fixed costs each month. Because all plant costs are considered fixed, this measure will tell management when the plant has covered its fixed costs and is beginning to make a monthly profit.

SUMMARY

The Brenham, Texas plant of Valmont Industries has significantly changed its operating methods over the past two years. Adopting the Theory of Constraints caused the Brenham plant management to reevaluate the way plant performance is measured. The traditional measures of efficiency, utilization, and standard cost did not provide management with adequate tools to evaluate the plant. The traditional accounting system remains in place and is used to report the operating results to headquarters. The figures that are reported to headquarters, however, are developed using some nontraditional assumptions.

Although component costs used to value inventory are developed using standard costing methods, the plant costs are all considered to be fixed. Variations, either favorable or unfavorable, are almost entirely eliminated. Exceptions are variations incurred because of changes in material costs or significant capacity changes in the plant.

The database that contains the routing and the standard hours is maintained and is utilized to schedule the noncritical resources. The pole assembly areas are considered to be the constraints. Performance measures are implemented in these areas as well. The organization has been changed as well; an individual supervisor is now responsible for the constraint. The second supervisor on the shift is responsible for the nonconstraint resources that support the buffers. The standard of performance for the constraint is 100% of scheduled hours. The standard of performance for the buffers is no holes in region 1 and 60% filled in region 2. The plant has found that if these criteria are satisfied, then orders will be shipped on time except when the outsourced galvanizing operations cause shipping delays.

Since adopting the principles of the Theory of Constraints as its method of operation, the Brenham plant has achieved significant improvements in its reported operating results. Total plant work-in-process inventories have been reduced by approximately 75% and, on-time shipments have increased from less than 85% prior to the implementation of the Theory of Constraints to in excess of 90% at present. In addition, the plant has been able to ship a significant number of orders early and offer some customers reduced lead times, giving the plant what management considers a competitive edge. The hourly work force has been reduced from 90 to 70; however, plant shipments have increased by 65% over the last two years. While figures are not available, changes in profits and return on the owner's equity have steadily increased in conjunction with these operating improvements.

APPLYING THE PERFORMANCE MEASUREMENT MODEL

Valmont/ALS found itself in a position similar to the many firms. The Brenham plant was in competition with two other Valmont locations for its business. Orders were allocated by Valmont's corporate production control function based on each competing plant's ability to meet customer delivery requirements. Long production runs were expected when equipment utilization and labor efficiency were the primary performance measures. These measures did not contribute to short lead times or improved delivery performance. The plant embraced the Theory of Constraints and developed a performance measurement system that assessed buffer status. In addition, the plant changed its cost measurement system to a system that identified opportunities for profitable business. Figure 7.2 summarizes the modifications the Brenham plant made to its competitive factors and its performance measurement system.

Competitive Factors	Lead time reduction and schedule performance— These factors support the corporate goals of continuously increasing profits and return on stockholder equity. Constraint and buffer operation are functions measured.
Constraint Performance Measure	Actual hours of operation versus scheduled hours.
	Standard hours recorded versus hours scheduled.
	Metric for both measures is percentage of scheduled hours.
	Performance standard is 100% in both cases.
Buffer Performance Measure	Region 1—All uncompleted hours on buffer work centers are individually measured and reported. The standard is zero uncompleted hours. If the total uncompleted hours reaches 20, expediting is initiated.
	Region 2—Actual hours completed versus hours required is measured. The metric is percentage completed. The performance standard is 50%. Values below 50% completed initiate expediting.
	Region 3—Not measured.
Shipping Buffer	Buffer contents are not measured. A two-week window provides protection against fluctuations in activities between the constraint and the shipping area.
Throughput Measurement	Measured and reported daily from shipping records. Used as an information and "scorekeeping" figure for managers, supervisors, and operating personnel.

Figure 7.2   Model Analysis Summary, Valmont/ALS

REFERENCES

1. Wahlers, J.L., A Study of Performance Measures in a Synchronous Intermittent Manufacturing Environment. Unpublished doctoral dissertation, University of Georgia, 1993.
2. Goldratt, E.M. and J. Cox, *The Goal: A Process of Continuous Improvement,* 2nd edition, Croton-on-Hudson, NY: North River Press, 1994.
3. Goldratt, E.M. and R.E. Fox, *The Race,* Croton-on-Hudson, NY: North River Press, 1986.
4. Johnson, H.T. and R.S. Kaplan, *Relevance Lost.* Boston: Harvard Business School Press, 1987.

THE TRANE COMPANY, INC., MACON, GEORGIA

BACKGROUND

The Trane Company is a large industrial organization headquartered in LaCrosse, Wisconsin. There are two industrial divisions, the Unitary Systems Group, which has seven factories and operates in a centralized manner, and the Commercial Systems Group. The Commercial Systems Group has 11 factories and operates with each factory as a strategic business unit. The Commercial Systems' factory in Macon, Georgia produces the company's line of large self-contained air conditioner units for buildings. The factory is about 15 years old and was originally built by Union Switch and Signal Company, another division of the American Standard Company. In 1987, the Union Switch product line was consolidated in another factory and the self-contained systems product line relocated from LaCrosse to Macon. The American Standard Company purchased Trane in 1984 and took the company private in a leveraged buyout in 1989. The Macon factory has recently received a second product line from a plant in Pennsylvania. This second product line, a smaller self-contained air conditioner unit, is produced in a separate portion of the factory. This study focuses on the larger product line, although the two product lines are managed in the same manner. In 1991, two additional products were added to the Macon factory, a refrigerant cleaning service unit and a refrigerant evacuation unit. These two products are produced in a separate portion of the facility and are considerably smaller physically than the self-contained unit. These latter two products will be excluded from the case.

The Macon plant's largest product line is air conditioner units that range from 20 to 80 tons. These units are produced to design specifications developed by a building contractor or engineer to meet a specific building's requirement. The product is assembled to order from more than 5000 components and subassemblies to meet the design specifications. There are four general product models, although over nine million unique designs are available according to management. The general product is about seven feet high, five feet wide, and ten feet deep.

The Macon factory operates on an eight-hour-per-day, five-day-per-week basis. Total plant employment is about 150 employees. The Macon plant uses both full-time and temporary employees in the production process. There are currently about 90 direct employees and 60 support employees. All permanent Trane employees are salaried. Temporary employees are hired from a local company as required to support production levels. These temporary employees are employees of the separate temporary employment company and receive no benefits from the Trane Company. Sometimes temporary employees are hired as permanent Trane salaried employees when vacancies exist. There is an unwritten understanding that permanent employees will not be subject to layoffs. The factory had annual sales of about $25 million.

Production Operations

The factory is arranged along an assembly line for the self-contained air conditioner unit (see Figure 8.1). Production begins with production of the air conditioner base. This is a steel frame that is assembled in a cube. The base travels down a roller conveyor line where the condenser and water piping are attached. A subassembly line produces the air conditioner coils and feeds the coil assembly into the frame, where it is mounted along with the condenser. An in-line test is conducted to identify leaks in the coil system. The unit then is charged with Freon and additional subassemblies are attached. The next major steps are the installation of wiring from the unit to the control panel and fan assembly. A second in-line test is conducted to assure proper wiring and unit operation. The side panels are then attached along with the outer doors, and the unit is painted and packaged for shipment. The assembly line is divided into 43 operations for work assignments grouped into nine work teams.

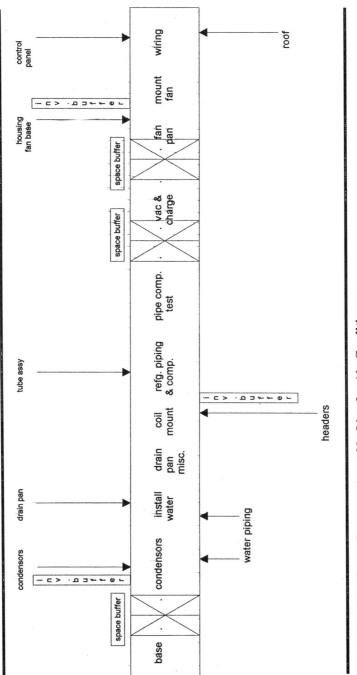

Figure 8.1   The Trane Company Assembly Line for Air Conditioners

The assembly line was designed as the product line was transferred into the existing facility in 1988. Because the factory was already in place at the time of the transfer, some compromises were made to fit the process into the building. Management planned to redo the factory layout in 1992.

Production Planning Environment

The Macon factory uses a team concept for the production organization. The operations manager is the chief operating officer for material control, manufacturing, and other plant operations. The operations manager reports to the factory's general manager. Reporting to the operations manager are three cell managers, a facilities manager, and a material/production control specialist. The cell managers act as the key coordinators for the respective cells rather than acting in the traditional managerial role. Production teams in each cell operate their cell as if it were a separate operating unit. Each cell has a team representative who acts as the intracell coordinator. The team representative has no actual production tasks to perform and, instead, facilitates the production decisions that are made to support the cell's overall production. The team representative position is rotated through the members of the cell every four months. Operating decisions are made by consensus of cell members. Staff operations, such as order entry, purchasing, and accounting, are not part of the cell organization but operate to support the manufacturing cells.

Production planning is a shared responsibility among the cell managers, the master scheduler, and the supplier development manager. Production decisions are made by consensus among the managers and the production employees. The materials/production manager is responsible for the physical movement of materials from material receipt through shipping. The materials/production control specialist is also responsible for cycle counting, inventory control, and inventory balance adjustments. The supplier development manager, who also reports to the general manager, is responsible for the inventory levels within the plant, as well as the normal purchasing functions such as scheduling, price negotiations, and vendor performance evaluations. The master production scheduler reports to the purchasing manager, as do three buyers, called material coordinators, and three supplier developers. The key performance measurement for the material/production control area is the inventory accuracy level. There is no specific performance measurement used for the master production schedule function.

Key performance measurements for the manufacturing cells are safety and housekeeping, delivery performance, inventory accuracy, quality, and productivity. The productivity measurement is based on a comparison of the total actual hours per unit and the planned hours per unit expected. The planned hours per unit expected is calculated from an equivalent unit-per-day production rate. Currently, the largest 80-ton unit has a 1.3 equivalent unit, while the smallest 20-ton unit has a 0.8 equivalent rate. The current master production schedule requires a 20-equivalent-unit-per-week rate. The actual production scheduled for a week is converted into the equivalent units for each cell area. The equivalent units are then converted into the number of production technicians required and then multiplied by the demonstrated productivity rate. The demonstrated productivity rate is the average of the previous four weeks of production compared to the planned production. The result of the demonstrated productivity calculation is used to adjust the planned number of production technicians required for each cell area. Additional hours are added for training and prototype builds, and the planned hours are determined.

Inventory management is also a shared responsibility among the cell managers, material/production control specialist, and the supplier development manager. Manufacturing lead time is two to five days. Raw material is 70% of the total inventory and has a current turnover of six times per year. Work-in-process is about 10% of the total inventory investment and has a turnover of 100 times per year. Finished goods inventory is about 20% of the total inventory investment and has a turnover of about ten times per year. Direct labor is about 10% of the total sales dollars. Material is about 60% of the total sales dollars. Overhead is about 30% of the total sales dollars. The key performance measurement for inventory management is the turnover rate. The annual targets were for raw material to turn 12 times per year, work-in-process inventory to turn 250 times, and finished goods to turn 15 times per year.

The marketing and sales functions are located at the Macon factory. There are about 2000 sales engineers located throughout the United States and overseas. The sales engineers work with building contractors and engineers to develop the air conditioning unit specifications to fit a particular work site. Units are engineered to order based on four general models and a variety of options and attachments. About 90% of all orders can be developed from existing options and attachments. About 10% of the orders require specific engineering design modifications, and both Trane design engineers and the sales engineers help the contractor create the modifications. There are four marketing engineers located at the Macon factory to facilitate the special requirements.

The marketing department at the Macon factory also has traffic responsibilities and the order entry function. Orders are entered via a computerized system that allows the sales engineers to select the base unit model, attachments, and options to create a salable configuration based on the air conditioner unit specifications developed at the building site. The order entry configuration uses the bill-of-material file to create a specific order number which defines each specific customer unit(s) on order. The typical customer order is for three individual air conditioner units, although as many as 100 units of the same configuration have been ordered. There are also three customer service representatives at Macon who monitor production and coordinate shipping with the customer to the work site. Key performance measurements for marketing are the day's billing outstanding, customer satisfaction, and sales volume.

The product development function is also located at the Macon plant. An engineering group of five engineers and six technicians is headed by an engineering manager who reports to the factory general manager. Product development also has an indirect relationship to the design engineering department at the Trane corporate headquarters in LaCrosse, Wisconsin. A new product line was planned for a 1992 introduction. As a result of the new product line, there would be a redesign of the factory layout and rearrangement of the assembly line tasks. Field service reports to the Macon engineering manager. Field service is generally the responsibility of the independent Trane dealer organization, although problems with a new installation would be investigated by the Macon field service representative.

There is no formal quality control or quality assurance department at the Macon factory. A total quality management program is in place, and quality is viewed as the responsibility of all employees. There are no inspectors employed at the factory. Quality is designed into the product and is verified in the production process by operators. There are two major test areas, one for leak testing and the second for checking overall unit performance, in the assembly line. These test areas permit corrections to be made at the point of assembly. Quality control graphs are located throughout the facility to provide feedback to operators. The method sheets that are used for assembly instructions also are color coded to identify inspection and verification activities for operators. Statistical quality control charts are not used at the Macon factory. Management states that no statistical process control is being used now "because no constraint to the business needs statistical process control at this time."

The accounting function consists of the factory controller, a cost accountant, a general accountant, and an accounting clerk. The factory uses a dual accounting system. External reports are generated to the Trane headquarters using traditional cost accounting methods based on direct labor hours. Profit and loss calculations are based on the cost accounting system. Internal factory management reports use the Theory of Constraints (TOC) approach to throughput, inventory, and operating expense. Management uses the TOC approach to evaluate changes to the manufacturing environment. Key performance measurements reported outside the factory are operating expenses before income and taxes, cash flow, inventory turnover, and outstanding days billing. Internal performance measurements include throughput, inventory dollar days, and operating expense.

The plant engineering function is a shared responsibility. About 80% of the manufacturing engineering functions are the responsibility of the manufacturing cell teams. A team coordinated by a facilities engineer works with the manufacturing cell teams to resolve problems. Two manufacturing engineers who report to the operations manager are responsible for any project activities. The two engineers are currently involved in the introduction of the new product line and the preliminary revised layout. Two engineering technicians who report to the operations manager are responsible for the coordination of engineering changes.

Logical Structure

Based on interviews and observations, the following characteristics are present at the Macon factory:

1. A relatively small number of components, about 3000 part numbers, are used to produce a relatively large number of end items.
2. The product flows through assembly points that are points of product divergence, where the unique end item emerges.
3. Several common manufactured and purchased components are assembled into a final end item.
4. Component parts tend to be common to many end items, while the assembly configuration determines the end item.
5. Routings for different manufactured components are quite dissimilar to one another, although component routings are assigned to specific workstations.

These factors lead to the conclusion that this plant can best be described as a T logical structure.

Additionally, characteristics tend to indicate that this plant is a repetitive manufacturer. The plant is described by management as repetitive, and the plant uses an assemble-to-order approach for production. The plant is arranged in a product layout, with one area dedicated to the large self-contained air conditioner unit and another dedicated to the smaller self-contained unit. Routings for all components are fixed. Production and capacity are measured in terms of a daily rate of production. The manufacturing lead time is relatively short, and changes to customer orders can be accommodated. Finally, the overall production process creates a flow of material through the assembly areas.

PRODUCTION PLANNING AND CONTROL METHODS

Background

The production planning and control system used at the Macon factory is a combination of JIT, TOC, and MRP. A MRP system was purchased in early 1989. The software package is identified as CONTROL, produced by Cincom, Inc. The decision to purchase CONTROL was made primarily based on a survey of company needs compared to various software packages. The system consists of several generic modules which are linked together by customized interfaces. The Macon factory uses a master production scheduling module, the MRP calculation module, the accounting module, and the bill-of-material module. A separate purchasing system was implemented at the same time as the MRP system. No shop floor control module was implemented, although the CONTROL package did have a module available. The decision not to implement the shop floor control module was based on the factory's desire not to use labor reporting. Labor reporting was a transactions driver in the CONTROL shop floor control module. The order entry system that is used was developed in-house and is common among all Trane factories. The order entry system feeds the master production scheduling system and uses the bill-of-material module by using a customized interface.

Management believes that the MRP system provided a framework for the other production planning and control methods used at the factory. A rough-cut capacity planning method is used that relies on the MRP bill of material and master production schedule (MPS) data. MRP data are used to provide customer promise dates and to support the product costing calculation that is used for external financial reporting. Management believes that the MRP system is providing very useful data for the types of decisions that require the data. Examples of uses of the MRP data are in costing decisions, engineering change management, and purchasing.

During the MRP implementation, the Macon factory used an outside consultant to provide management education. The system required about 1000 man-hours to implement. There was no MRP training provided to the manufacturing cell employees except for the cell managers.

The company employed another outside consultant in 1988 to improve aspects of a JIT approach. The consultant used the term "demand flow management" to describe JIT concepts, and the factory uses the consultant's term more than it uses JIT. Management reports that it uses a leveled master production schedule, production schedule linearity, scheduled preventive maintenance, multiskilled employees, and made revisions to the plant layout to accommodate JIT. The Macon factory management reports that it does not currently have a setup reduction program and that one is not anticipated or "needed." The kanban method of material movement of subassemblies to the final assembly line is used. Suppliers also use a triggering system to replenish hardware and other purchased components. Cell managers have been trained to train employees in JIT methods.

The factory has implemented a TOC approach to production planning and control as well. The TOC approach was begun in 1989 and continues to be evolving. The performance measurement aspects of throughput, inventory, and operating expense are used for internal management decisions. The TOC five-step process is also used by management to provide a method of continuous improvement through the production process. Management also reports the use of the effect-cause-effect and clouds decision-making techniques. Finally, the shop floor is controlled by the use of the drum-buffer-rope and buffer management techniques (see Figure 8.2). The former general manager and the current operations manager have been educated in the TOC, as have other members of the management team. Management has no plans to use the TOC disaster software and has not used the OPT software.

The accuracy of the bill of material is reported to be above 97%. The bill of material is continuously undergoing changes to improve the product. Currently, there are about 100 engineering changes in various stages of implementation. Each customer order has a unique product configuration. The customer bill of material is called the sales order bill. The sales order bill has an accuracy of about 85 to 90% compared to the actual product that

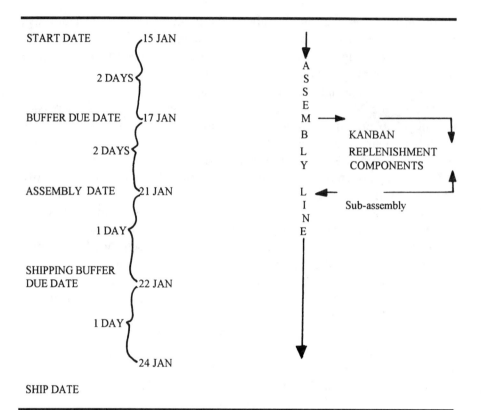

Figure 8.2   Kanban and Buffer Management Example (enough kanbans are in circulation to provide the two-day buffer from subassembly)

will be installed. The errors on the sales order bill occur as the result of human errors in creating the order. The most common errors involve the quantity per item required in the assembly process. An audit team is devoted to correcting the sales order bills and identifying areas where cost reduction and product improvement can be made. Errors on the sales order bill are often found and corrected during the assembly process. However, some errors cause inventory errors as a result of the MRP backflush consumption and cause purchasing errors and expediting. Currently, the factory has suffered about 35 lines stoppages as the result of parts shortages during the previous month. The peak line stoppage month had about 70 shortages. The current goal is less than ten line stoppages per month.

Management reports that the routings are 100% accurate. Routings are rarely changed, and when a change occurs, it is part of a larger product introduction program. Management believes that no problems are generated by the component routings.

Inventory accuracy is seen to be a key area of concern. There is a cycle counting function where class A parts are counted weekly, class B parts are counted every two weeks, and type C parts are counted monthly. A physical inventory is conducted on an annual basis as part of the outside auditing requirement. Management is allocating resources to improve the inventory accuracy, and visible performance charts are kept in view for employees to review. The MRP system uses a lot-for-lot order policy in weekly buckets for purchasing requirements. Safety stock varies by part number and is determined by the purchasing department and the suppliers. About 33% of all purchased part numbers have safety stock of about two weeks. The supplier development manager has three supplier developers who work with vendors to improve vendor performance so that the safety stock levels can be reduced. All safety stock is maintained in the raw material areas. Lot sizing for purchased components depends on the vendor's packaging and part characteristics. The goal of the purchasing department is to have as many suppliers use kanban triggers as possible. Currently, about 10% of purchased parts are managed using kanban.

Master Production Scheduling

The master production schedule (MPS) is a shared responsibility among the operations manager, purchasing manager, cell managers, and marketing manager. There is no interaction between the Macon factory and the Trane Company headquarters concerning the MPS. Four base models are used to develop the sales forecast and the MPS. Options and attachments are forecast as a percentage and the quantities placed in the MPS. The MPS is then fed into the MRP module for the bill-of-material explosion and MRP calculation.

A sales and operations plan is developed on an annual basis for the Macon factory. This sales and operations plan is approved by the Trane Company upper management and becomes the basis for the factory's budget. The sales and operations plan is based on a sales forecast developed by the Macon factory's marketing department. The forecast is based on historical sales, the expected demand, and modifications to the product line. The forecast is in monthly buckets for the next 12 months. The operations group, including the operations manager, cell managers, and purchasing manager, review the sales forecast and determine the resources necessary to support the build rates. The operations group, marketing, and plant manager review the sales and operations plan on a monthly basis and make any adjustments as required. The monthly forecast is broken down into weekly quantities by the master production scheduler, who reports to the purchasing manager.

As actual customer orders are received in the order entry department, the forecast quantity is replaced. The first four to six weeks are expected to be filled by actual orders in the master schedule. Actual orders can be placed at any time during the 12-month planning horizon. Since air conditioner units are larger items at a construction site, the building contractors and/or architects are able to give considerable visibility of the order to the master production scheduler. The factory tries to maintain the first two weeks in the MPS as a frozen time period where new customer orders are not placed. Changes to existing orders are reviewed and accepted if possible, however. The MPS feeds the MRP system on a daily basis. The MRP system is operated as a net change system for purchasing requirements.

The actual customer orders contain the configuration required to support a specific building requirement. The options and attachments required to build the actual order are extracted from the order and transmitted into the MPS automatically. These actual order quantities replace the forecasted quantities that had been derived from the percentages used in the sales forecast.

A production control specialist sequences the weekly production build by physically hanging build packets in order at the gating operation on the assembly line. The assembly line is currently scheduled for a 20-unit-per-week build rate. The sequence of the line build is based on the desired customer ship date from each customer order.

The factory uses a one-week shipping buffer which is supported by the master production scheduling dates used by the master scheduler to launch units at the gating operation. A two-day buffer is planned at the constraint in the assembly line. Actual manufacturing assembly time is two to three days. Therefore, the MPS executes the drum-buffer-rope method at the Macon factory. The drum is set at the constraint in the assembly line, which is known to the master production scheduler. The transmission of the build packets that establishes the build sequence is the rope. Finally, a two-day buffer exists at the constraint. The assembly line components that do not go through the constraint have an informal assembly buffer of one to two days as the result of the kanban minimum-maximum levels used and the purchasing system's weekly bucket. An overall five-day shipping buffer exists as part of the MPS calculation. There is no assigned buffer manager. Instead, the cell manager where the constraint is located is responsible for monitoring the buffer and taking action as required.

The execution of the MPS is the combined responsibility of the cell managers, purchasing manager, and all other employees. The actual build amount is posted on a visible display at each of the nine work areas and in the employee lunchroom. Any backlog is made up during the production week on an overtime basis.

Priority Planning

Priority planning for purchased parts is largely controlled by the MRP system. A sales order has a five-day shipping buffer created in the offset from the ship date to the build date. The MRP system has weekly buckets and cannot accept any time period of less than one week. For example, a customer ship date of 22 November will be placed in the MPS on 15 November. All purchased material is due on the assembly line 8 November, with a material arrival date of 4 November. In addition, about 10% of the purchased parts are planned using kanban triggers.

Priority planning for manufactured components is managed using the drum-buffer-rope technique. A buffer is planned at the assembly line constraint. Currently, the assembly line constraint is the Freon charging and vacuum task. A space buffer exists after the constraint as well. The constraint buffer is one day. Material flows into the assembly line from the feeder operations in accordance with the build packet. The build packet, discussed in the master production scheduling section, consists of the job name, options and attachments required, and the method sheet (assembly instructions). Kanban is used for subassemblies flowing into the final assembly line. All major subassemblies have visible minimum and maximum limits established by some physical means at the point of use. If the minimum limit is not reached, the feeding operations schedule overtime as required. If the maximum is reached, the feeding operations stop production.

The constraint buffer also establishes a minimum and maximum level of the major component. The cell manager monitors the buffer amount and adjusts the cell work plan as required either by expediting to increase the buffer or by using overtime to reduce the buffer. No documents are used on the shop floor to communicate planned part priorities.

Capacity Planning

Capacity at the Macon factory is defined in terms of equivalent units per day. The current capacity at Macon is six units per day. This capacity is defined on a one-shift basis, although additional shifts could be added if required. The factory's capacity is measured at the assembly line constraint. Capacity is not measured at any other machine center in the plant. The work force level is calculated from the daily production rate for each of the nine workstations. This calculation is made on the same worksheet as was described in the section on production planning background.

The Macon factory conducts long-term capacity planning based on a three-year sales forecast developed by marketing. The results of the long-term capacity analysis are forwarded to the Trane Company headquarters along with any requests for additional capital.

Rough-cut capacity planning occurs at the time the annual sales and operations plan is developed. Capacity is reviewed as an integrated part of the master production scheduling process. Actual customer orders replace the forecasted quantities in the MPS during the first several weeks. If actual orders do not replace the entire quantity in one week, customer orders are pulled into the available week. If the amount of actual orders is not sufficient to replace the forecast in the first four weeks, management begins to more closely monitor sales and may take extraordinary actions to generate additional sales. If additional sales cannot be added, an adjustment to the MPS will be made to lower the daily production rate.

If the actual customer orders are sufficient to replace the forecast in the MPS beyond six weeks, management may take action to increase the daily production schedule rate. The first action taken by management to add or subtract the daily production rate is the use of temporary employees. Approximately 35% of the total operations are designated as temporary jobs. If a capacity adjustment is required, the level of temporary employees is adjusted. Management has no plans to reduce the number of full-time production technicians under any circumstances. As more temporary workers are added, up to the one-fourth limit, management may decide to hire new full-time employees if the market appears to support a long-term addition to the permanent work force.

The factory uses the data from the MPS and the MRP calculation of production orders as a basis for reviewing the level of workers assigned to each workstation on the assembly line. This calculation is based on the forecasted option and attachment percentages in the MPS and the standard labor hours from the routings. A 12-month horizon is used for this calculation. This calculation is updated on a monthly basis as part of the master production scheduling process.

In the short run, if capacity exceeds the amount of actual customer orders, the number of temporary employees is reduced for the next day or week. If the amount of actual orders is greater than capacity or if there is a need to catch up to the production scheduled or to replenish a subassembly that has fallen below the kanban minimums, the employees at the affected work cell schedule overtime as they determine it is needed. The cell manager coordinates the cell's scheduled overtime with other cell managers. No management approval is required to schedule overtime at the cells.

Priority Control

Each cell team assigns one employee to be a team representative on a rotating basis for four months. The team representative has no physical production tasks assigned. The primary duty of the team representative is to coordinate information with the other team representatives concerning production requirements and material availability. On a daily basis, the production control specialist reconciles the actual build amount with the MPS. This information is given to each team representative and to accounting for the monthly production reports. The team representatives share the production information with other team members and together develop a plan to take any necessary action required to keep on schedule. Neither a daily dispatch list nor a manufactured shortage report is used. If there is a problem at the constraint buffer, the team managers may also become involved in creating the action plan, as might the purchasing manager if purchased part shortages are involved. The constraint buffer is continuously monitored by the team representative where the vacuum pump operation is located.

A shortage list is managed for purchased parts shortages. There are three buyers who also act as material coordinators as required to expedite purchased material. The MRP system is the basis for parts shortage notifications that are printed each morning. The shortages are the result of the previous night's inventory backflush consumption calculation and the week's build schedule amount. As a result of this calculation, the buyer might receive a shortage notification with very little lead time to react. Consequently, there are purchased parts shortages that impact the build schedule. The effect of the shortages is to stop the assembly line. The line stoppages have averaged about 35 per month during the past several months. The line stoppages become part of the work force calculation, as the stoppages impact the productivity factor calculation. Management is devoting considerable attention to reducing the number of line stoppages. The current goal is to reduce the number of stoppages to less than ten per month. Management believes the largest contributor to the line stoppages is errors that occur in the sales order configuration which result in an inventory error.

Capacity Control

As discussed in the capacity planning section, the Macon factory relies on the use of temporary workers for short-term capacity adjustments. The temporary workers are employees of an independent temporary employment agency that is located in the Macon area. On a daily basis, the Macon factory's human resource department contacts the temporary agency and hires the number of workers required for the next day or week. There is no expectation that a temporary worker will be hired as a permanent employee. Temporary workers are assigned to production tasks that are labor intensive and require little training.

On a daily basis, the reconciliation of the actual build to the build schedule discussed in the priority control section is conducted. Based on the daily reconciliation, the cell teams calculate the number of temporary workers that are required for the next week. This information is passed to the human resource department for action. On a weekly basis, each cell manager coordinates with other cell managers and the operations manager on the number of temporary workers used during the week and the number expected to be used the following week. This coordination may lead to the addition of new permanent employees and an adjustment to the MPS build rate.

KEY PROBLEM AREAS BY PRODUCTION FUNCTION

Master Production Scheduling

The first area of concern is the seasonality in demand for the final product. Building construction is seasonal in many parts of the country. The MPS is developed with the expectation that a leveled production rate can be produced over a relatively long period of time. The more stable the master production schedule, the more stable the purchased parts schedule for vendors. The peak tends to occur at the end of the summer and during the fall time period, while the valley tends to be in the winter and early spring.

A second concern is the coordination difficulty between the master production scheduler and the marketing department if there are insufficient customer orders to fill the production schedule. The coordination is difficult because it involves generating sales from the field and may involve pricing decisions.

A third concern is the limits in the MRP software package. The MPS is defined in weekly time periods and shows all requirements due on Friday, even though production is scheduled to occur daily. As a result, inventory is purchased earlier than needed, and communication difficulties occur between purchasing representatives and suppliers when requirements shift from one week to another. A fourth concern is the relatively long lead time required for certain purchased components.

A fifth area of concern is the use of forecasted options based on percentages in the MPS system. Because of the complexity in the design of the air conditioning unit, a large number of options and attachments are required for each customer order. The forecasted quantity may not be the same amount as has been forecast and ordered through the purchasing system. The result is that expediting is required when an actual customer order is configured differently than the quantities of component parts that are on order from vendors.

Priority Planning

A key area of concern is the limitations in the MRP system as it calculates requirements for purchased parts based on weekly time periods. A small inventory adjustment may cause the movement of an entire purchased part order from one weekly bucket to another. As a result, purchasing may be required to expedite purchased parts. If the expediting is not effective, a line stoppage can occur, impacting the entire factory.

A second concern is the lack of inventory accuracy in the MRP system for purchased components. The result of inaccurate inventory records can be expediting purchased parts, as described above.

A third area of concern involving priority planning is the management of the production constraint. The production constraint was originally the coil subassembly which was located at the first operation in the assembly line. Coil equipment was purchased based on TOC calculations. As a result, a new constraint emerged at the vacuum pump operation. Consequently, the buffer had to be relocated and the entire assembly line is now viewed as two separate areas for cell management attention. The result of the new constraint is that the entire shop floor method had to be reviewed.

A fourth area of concern is the calculation of the size of the constraint buffer. Since the physical assembly is relatively quite large, the amount of space required for the constraint buffer and the space buffer had to be determined and changes made to the existing assembly line. Management was concerned that a miscalculation of the buffer size would have serious results that might take considerable time to resolve. There was little documentation available concerning the calculation of the buffer size in TOC literature.

Capacity Planning

The first concern for capacity planning is the difficulty in developing a valid sales forecast. The lead time for new capital acquisition is relatively long and, therefore, requires a valid forecast. The economic recession in the commercial building industry has negatively impacted the Macon factory and made the forecasting process more difficult. Management believes that an assembly line operation requires considerable lead time to alter long-term capacity, as a revised layout and reassignment of tasks may be required.

A second concern for capacity planning is the difficulty in identifying nonproductive time requirements for the work force, such as training time, prototype builds, and illness. The nonproductive time is used in the productivity factor calculation and impacts the number of temporary employees hired.

A third area is the location of the constraint in the middle of the assembly line rather than at the gating operation. Management feels that there is a lack of timely information about the constraint in order to keep a proper amount in the buffer.

A final area of capacity planning identified by management is the lack of a finished goods storage area. The addition of new products to the facility required the relocation of finished goods and an overall reduction of floor space. Macon managers feel a reduced cycle time for the products is needed to eliminate the extra stock.

Priority Control

The major concern for priority control is the amount of expediting required for purchased components as the result of inventory errors. Expediting is disruptive to the production process and runs counter to management's emphasis on planning. As a result of the inventory errors, there have been a significant number of line stoppages during the previous months. The inventory errors and resulting expediting are having a negative impact on relations with some vendors and the JIT program.

A second concern is the adjustment of the build sheets as the result of human errors in the ordering process. The errors in the build instructions can cause inventory errors when the backflush inventory consumption occurs based on the order bill rather than the actual build.

A third concern is the lack of part information on the build sheets except for the options required at the workstation. Management believes that additional inventory errors occur when workers must remember or guess the components required at the workstation.

Capacity Control

The key concern in capacity control is the difficulty in predicting the amount of nonproductive time required during the week. The lack of information results in the lack of enough workers to perform the assigned production requirements. The result may be unnecessary overtime. The other concern is the amount of time required to train temporary workers for some tasks. Training takes time away from other productive activities and also may cause unnecessary overtime.

WITHIN-CASE ANALYSIS

Data gathered from the previsit survey and on-site interviews identified several areas of concern for the Macon factory management. Table 8.1 summarizes the methods used to plan and control production at the factory. Table 8.2 summarizes the key areas of concern in each of the production planning and control functions. Table 8.3 identifies the critical characteristics of the production planning and control system and the impacts of the characteristics on the logical structure.

CRITICAL CHARACTERISTICS AND IMPACTS BY FUNCTION

Master Production Scheduling

The first key characteristic of the master production scheduling function is the instability of the sales forecast. The factory forecasts using four base air

Table 8.1   Production Planning and Control System Methods Used at Trane Company, Macon, Georgia

Master Production Scheduling

1. Sales forecasting and managerial judgment using a team approach are used as the basis for developing the MPS.
2. The MPS is developed to maintain a level production schedule for the assembly line and is reviewed on a monthly basis.
3. Actual orders replace forecasted quantities during the first four to six weeks.
4. A final assembly schedule is prepared from the MPS and a build schedule is established by hanging build packets at the final assembly operation.
5. Actual production is monitored on a daily basis and reported to the work force to allow for any adjustments to the work force.

Priority Planning

1. The factory has a product layout along an assembly line with feeder subassembly operations.
2. The manufacturing of subassemblies is based on a minimum/maximum level established by physical restrictions and/or kanbans.
3. A one-day constraint buffer is used to protect the assembly line constraint. A space buffer is also used to prevent constraint stoppages from blockages.
4. The material flows down the assembly line according to the build packet that is the final assembly scheduling tool.
5. Purchased parts are ordered from MRP data that are fed from the MPS actual orders and forecast. MRP is limited to weekly time buckets.
6. The purchased parts inventory is consumed by a backflush in MRP at the point of final product build.
7. About 10% of all purchased components are managed by the kanban method.

Capacity Planning

1. Long-term capacity is planned on an annual basis, along with the sales and operations plan that is forwarded to Trane headquarters for capital acquisitions.
2. Medium-range (one year) capacity is planned on an annual basis and updated each month as an integrated part of the master production scheduling process.
3. The factory uses a relatively large number, up to 25%, of temporary workers on a continuous basis for production tasks.
4. Adjustments to planned capacity are made by the addition or subtraction of the temporary work force. Changes are made to the full-time work force only when the change appears to be permanent in the market for end products.
5. Both short-term and medium-term capacity are adjusted by management of the temporary work force.

Priority Control

1. Manufacturing parts are controlled by the use of the drum-buffer-rope method and by buffer management techniques.

Table 8.1   Production Planning and Control System Methods Used at Trane Company, Macon, Georgia (continued)

2. Levels of manufactured subassemblies that fall below preestablished minimums trigger manufacturing actions to add components. Levels that would be beyond the maximum limits trigger manufacturing actions to stop production.
3. Purchased parts priorities are controlled by NW reports, including the purchase order, a shortage list, and informal communication. Expediting is undertaken as required to maintain production.

Capacity Control
1. The daily production status report is distributed to the manufacturing team representatives.
2. The team representatives communicate with one another and, with their cell teams, create a work force plan to make capacity adjustments as needed to support the build schedule.
3. If overtime is the chosen method, the work teams schedule the overtime and number of employees needed. Workers are multiskilled and salaried.
4. If addition or reduction in the number of temporary employees is the chosen method, the work team representative forwards the status report to the factory's human resource department, where arrangements are made with the local employment agency for an adjustment the following week.

conditioner models in the 20- to 80-ton range and approximately 2000 options and attachments. The options and attachments are forecast based on the historical percentages that were constructed during the previous year. The second key characteristic is the relatively large number of salable configurations that must be manufactured during the year. The unstable forecast and the large number of end items impact the master scheduling function by causing management to select a make-to-order production planning environment and a relatively short (four to six weeks) horizon of actual customer orders from which to develop the MPS.

The third key characteristic is the limitation imposed on the planning process by the MPS module in the MRP system used at the Macon factory. The MPS module supports only weekly time buckets, so that all of a single week's production requirements are scheduled to be built on each Monday, even though the factory actually schedules production to be assembled each day. As a result, the factory management makes relatively limited use of any capacity analysis tools that could normally be used in developing the MPS.

The fourth characteristic is the use of the team concept at Macon. The

Table 8.2   Summary of Key Problem Areas at Trane Company, Macon, Georgia

Master Production Scheduling
1. There is seasonality in the final product demand.
2. Forecasts are unstable because of the recession. Marketing sometimes cannot fill the desired six weeks with actual orders.
3. The weekly time bucket in the MPS system drives the MRP system with a lump sum demand on each Monday rather than an even schedule each day.
4. The complexity in the product design and options required is difficult to forecast using a historical percentage.

Priority Planning
1. The MRP system uses weekly buckets for all parts even though the plan is to build some products each day.
2. Inventory accuracy has been a problem.
3. Purchased parts shortages have caused line stoppages about 35 times per month.
4. The constraint buffer size was difficult to calculate and causes some uneasiness for management.

Capacity Planning
1. The forecast inaccuracy causes capacity calculations to be inaccurate.
2. Identification of nonproductive times is difficult to forecast and can adversely impact the productivity factor used in capacity calculations.
3. The constraint's location in the middle of the assembly line causes some to view the line as two separate entities.

Priority Control
1. Parts are not identified on the assembly packet except for options, and human errors can cause inaccurate inventory and production errors.
2. Inventory errors cause expediting for purchased components.
3. Errors can occur in the sales build order as the result of human errors, which cause inventory errors and problems in assembly.

Capacity Control
1. Nonproductive time is difficult to forecast and causes overtime.
2. Training for temporary employees takes away from productive time.

MPS is developed and reviewed on a monthly basis by the operations team. The decisions necessary for MPS development are made using a consensus approach with participation by the cell managers, facilities managers, manager of purchasing, and the operations manager. The fifth characteristic is the use of a product layout, specifically an assembly line, to orchestrate production. The team approach plus the existence of an assembly line layout combine with the limitations of the MPS module to cause management to integrate the capacity planning and MPS functions. This is simplified by Gantt charting the constraint and then basing the line scheduling on the constraint capacity.

The sixth characteristic is the use of short fixed routings for similar

Table 8.3   Critical Characteristics at Trane Company, Macon, Georgia

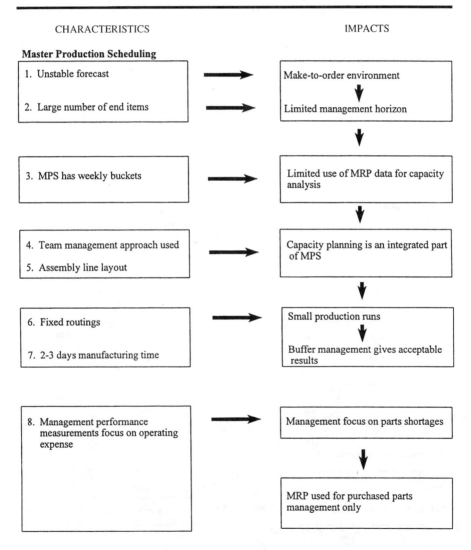

CHARACTERISTICS

IMPACTS

Master Production Scheduling

1. Unstable forecast

2. Large number of end items

Make-to-order environment

Limited management horizon

3. MPS has weekly buckets

Limited use of MRP data for capacity analysis

4. Team management approach used

5. Assembly line layout

Capacity planning is an integrated part of MPS

6. Fixed routings

7. 2-3 days manufacturing time

Small production runs

Buffer management gives acceptable results

8. Management performance measurements focus on operating expense

Management focus on parts shortages

MRP used for purchased parts management only

Table 8.3   Critical Characteristics at Trane Company, Macon, Georgia (continued)

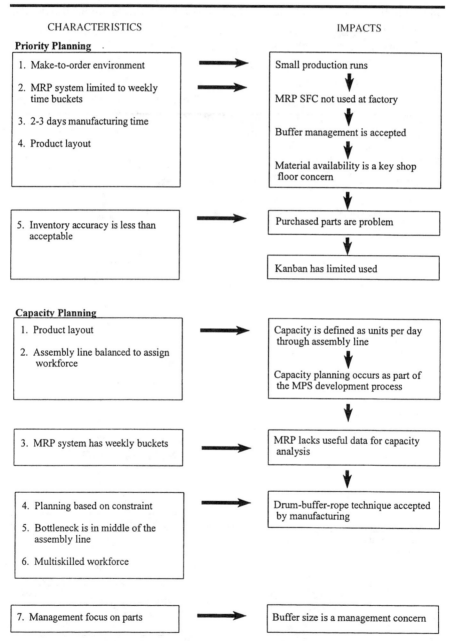

CHARACTERISTICS   IMPACTS

Priority Planning

1. Make-to-order environment   ⟶   Small production runs

2. MRP system limited to weekly   MRP SFC not used at factory
   time buckets

3. 2-3 days manufacturing time   Buffer management is accepted

4. Product layout   Material availability is a key shop
                                                   floor concern

5. Inventory accuracy is less than   ⟶   Purchased parts are problem
   acceptable

                                                   Kanban has limited used

Capacity Planning

1. Product layout   ⟶   Capacity is defined as units per day
                                                   through assembly line
2. Assembly line balanced to assign
   workforce   Capacity planning occurs as part of
                                                   the MPS development process

3. MRP system has weekly buckets   ⟶   MRP lacks useful data for capacity
                                                   analysis

4. Planning based on constraint   ⟶   Drum-buffer-rope technique accepted
                                                   by manufacturing
5. Bottleneck is in middle of the
   assembly line

6. Multiskilled workforce

7. Management focus on parts   ⟶   Buffer size is a management concern

Table 8.3   Critical Characteristics at Trane Company, Macon, Georgia (continued)

CHARACTERISTICS	IMPACTS

Priority Control

1. Product layout/assembly line →	Priority established by the final assembly schedule
2. MRP system limitations →	MRP SFC module not used
3. Location of bottlneck 4. Three-day manufacturing time →	Drum-buffer-rope environment accepted by manufacturing
5. Inventory problems	Material availability is a key area of concern

Capacity Control

1. Product layout/assembly line used →	Capacity status is determined by line build status
2. MRP system limitations →	MRP data not useful
3. Focus on buffer status →	Buffer management technique accepted by factory
4. Unplanned non-productive time 5. Buffer maintains constraint effectiveness 6. Team approach used →	Use of overtime and temporary workforce adjustments are primary tools

components in the subassembly and assembly areas. The seventh characteristic is the relatively short actual production time for assembling the end
item. The short fixed routing structure and the short production times combine to cause the factory to produce in relatively small lots. The use of small
production runs facilitates the use of the drum-buffer-rope technique by
providing a flexible environment where the process batch can be easily
separated from the transfer batch. This, in turn, allows positive results from
the buffer management technique of shop floor control.

The eighth characteristic is the dual performance measurement system
used by the Macon factory management. The traditional cost system is used
to report externally, while the TOC system is used for internal decision
making. As a result, management tends to focus on the operating expense
performance measurement. This has focused management's attention on the
parts shortages that have caused the assembly line stoppages to occur. This
focus creates an environment where the MPS system is used primarily for the
generation of purchased part requirements through the MRP calculation.

Priority Planning

The first key characteristic of the priority planning function is the make-to-
order planning environment. The second characteristic is the limitations of
the MRP system used at the Macon factory. The MRP system has weekly
time buckets which cannot support production quantities used on a daily
basis. However, the assembly line operates to produce units on a daily basis.
Further, the third characteristic is the relatively short production time required. The production time for the complete assembly of an air conditioner
unit is less than one week. Therefore, a conflict occurs between the MRP
requirements and the actual production of the air conditioner unit. The fourth
characteristic is the use of an assembly line for production of the units. The
characteristics of the make-to-order environment, the MRP system's limitations, the short production time, and the use of an assembly line combine to
impact the priority planning function by creating relatively small production
runs. Further, the four characteristics also combine to create a production
environment where no formal MRP shop floor control methods can be used.

An additional impact of the characteristics described above is that the
small production runs and the lack of a formal MRP shop floor control
approach have created an environment where the drum-buffer-rope and buffer
management techniques were well accepted by the manufacturing and management teams. However, the use of buffer management also focuses

management's attention on the availability of material to support planned production. The fifth characteristic is the lack of inventory record accuracy in the MRP system. The lack of inventory accuracy causes purchased parts to be expedited to support planned production. As a result of the expediting, there has been relatively limited use of kanban control methods for purchased parts.

Capacity Planning

The first critical characteristic of capacity planning is the product layout of the factory. This characteristic combines with the second characteristic, the factory's use of line balancing to assign the work force to tasks, to impact the capacity planning function so that capacity is defined as units per day. Further, defining capacity as units per day and the product layout and use of line balancing lead to integrating capacity planning with the development of the master production scheduling function.

The third characteristic is the MRP system limitation of supporting only weekly time buckets. The limitations of the MRP system lead to the lack of MRP data being used for capacity planning during the MPS process. The lack of useful MRP information for capacity planning created a supportive environment for the use of the drum-buffer-rope technique by both manufacturing and management teams.

The fourth characteristic is that capacity planning is based on the constraint. The fifth characteristic is the existence of the constraint in the middle of the assembly line. The sixth characteristic is the existence of a flexible work force. The seventh characteristic is the focus of management attention on parts shortages that cause stoppages in the assembly line. The existence of the constraint and the attention to shortages lead to management's concern about the size of the constraint buffer being used.

Priority Control

The existence of the product layout in the form of an assembly line and feeder operations is the first key characteristic of priority control. The second key characteristic is the limitations of the MRP system. The MRP system supports only weekly buckets and conflicts with the execution of the final assembly schedule which plans for daily production builds. The assembly line and the lack of MRP data cause the priority for manufacturing components to be determined by the final assembly schedule. The final assembly schedule is executed by arranging the build packets in a particular order at the first assembly operation. As a result of the use of the final assembly schedule, the manufacturing organization can effectively execute production without MRP's shop floor control tools. This lack of MRP shop floor control tools creates a favorable environment for the acceptance and use of the drum-buffer-rope method.

The third characteristic of the priority control function is the location of the constraint within the assembly line. The fourth characteristic is the relatively short production time required to assemble a unit. The existence of the constraint in the assembly line and the short assembly time also impact the priority control function by supporting the drum-buffer-rope technique. Both manufacturing and the management teams accept the results of the drum-buffer-rope technique and adjust production accordingly.

The last characteristic is the existence of inventory problems at the factory. This characteristic impacts the priority control function by focusing management's attention on material availability as a critical area of concern. The focus of attention results in the use of MRP for purchased parts scheduling and execution.

Capacity Control

The first key characteristic of the capacity control function is the existence of the assembly line. The assembly line impacts the capacity control function by causing the capacity to be determined by the status of the build rate and the final assembly schedule.

The second characteristic is the limitations of the MRP system. The MRP system supports only weekly buckets cannot be used to create intermediate start dates and due dates for production requirements. The lack of valid MRP planned orders impacts the capacity control function in that MRP reports are not useful in controlling capacity.

The third characteristic is the focus on the buffer status of the drum-buffer-rope technique. The lack of valid MRP information and the successful results in monitoring and adjusting production based on the buffer status cause a favorable environment for the use of buffer management to exist at the factory in both manufacturing and management teams.

The fourth characteristic is the existence of unplanned nonproductive time that adversely affects the cell work teams. The fifth characteristic is that the buffer maintains constraint effectiveness. The sixth characteristic is the use of cell teams to manage the production environment. The unplanned nonproductive time, the use of a team consensus approach, and the effectiveness of the buffer means that overtime and adjustment of the temporary work force are the primary capacity control tools.

SUMMARY

The Macon factory of the Trane Company uses a variety of production planning and control methods, including MRP, TOC, and JIT. The factory uses MRP primarily for the development of the master production schedule, for the bill of material, and for generating purchasing requirements. The TOC performance measures is used primarily for internal management decision making and drum-buffer-rope and buffer management for shop floor planning and control. The JIT methods are used primarily for facilitating the actual movement of material. Management feels that the use of multiple methods of production planning and control allows the factory to maintain a competitive advantage and has been rewarded by the assignment of new products to the Macon factory.

THE STANLEY FURNITURE COMPANY, STANLEYTOWN, VIRGINIA

BACKGROUND

The Stanley Furniture Company is a manufacturer of a broad line of wood furniture including bedroom suites, dining room suites, occasional pieces, home office furnishings, and entertainment centers. Stanley Furniture was founded in 1924 through the merger of two other furniture companies with roots in the mid-1800s. The company was acquired by Mead in 1969 to form the Mead Furnishings Group. Since 1979, the Stanley Company has been involved in five separate leveraged buyouts and sales. In 1989, Stanley Furniture was taken private by its current owner.

The Stanley Furniture Company is headquartered in Stanleytown, Virginia and has five factories. The largest factory is located next to the company headquarters in Stanleytown. Other factories are located in West End, Robbinville, and Lexington, North Carolina and in Waynesboro, Virginia. Only the Stanleytown factory makes tables, chairs, and cabinets. The other smaller factories produce only cases, beds, and cabinets.

Stanley Furniture is a centralized operation with its factories operating as profit centers. Each factory has a vice-president of production who reports to the senior vice-president of manufacturing. The senior vice-president of manufacturing reports to the executive vice-president and chief operating officer of the company. Approximate annual sales for the company are $145 million. About 2500 hourly and 330 salaried employees work for the Stanley

Furniture Company. This case study discusses the Stanley company head-quarters and the Stanleytown, Virginia factory, which consists of seven separate organizational plants.

Production Operations

The Stanleytown factory operates on an eight-hour-per-day, five-day-per-week basis. The factory produces approximately 2000 end items or about one-third of the total end items produced by the company. Stanleytown production is focused on the manufacture of suites of furniture rather than individual pieces. A suite typically consists of about seven items for a dining room and five items for a bedroom. Great care is taken to match the wood pattern as well as the type of wood in the manufacture of suites as opposed to individual occasional items. There are approximately 160,000 part numbers, including subassemblies and components, at the Stanleytown factory. The typical bill of material is five levels deep. About 95% of the furniture produced at the Stanleytown factory is make-to-stock.

The Stanleytown factory employs 1066 hourly workers and 66 salaried employees. The Stanleytown factory is organized into seven plants (see Figure 9.1). Some plants are physically separated from others, whereas others are separated by organizational lines. Plant 1 produces case goods, including bedroom and dining room cabinets. Plant 2 produces occasional items, dining tables, and some bedroom furniture. Plant 3 produces chairs, mirror frames, ladders, and bunk bed rails. Plant 4 finishes all items except chairs. Plant 5 is the panel plant for the production of veneers. Plant 6 is the dimension mill. Plant 7 is designated as the high-technology area where ornate cuttings and shaping takes place. The following section describes the manufacturing process of a typical furniture suite in the order of manufacture rather than in plant numerical order (see Figure 9.2).

Raw material is purchased in bulk from lumber distributors. Raw material is delivered to the factory by truck in loads called "hacks." Each hack is a stack of rough lumber from the lumber mills that is approximately 20 feet long by 8 feet wide and 6 feet high. The raw material area at Stanley is six acres of hacks stacked about 30 feet high by type of wood. As hacks are delivered, the lumber is graded in a separate building and identified by the grade and delivery date. Lumber in the stacks undergoes a drying process depending on the moisture content of the wood and the type of wood species. Average drying time is about four months, although there is wide variation

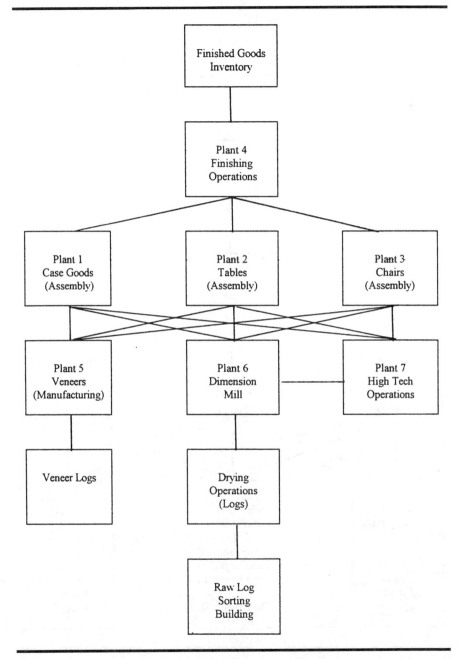

Figure 9.1 Stanleytown Factory Conceptual Framework

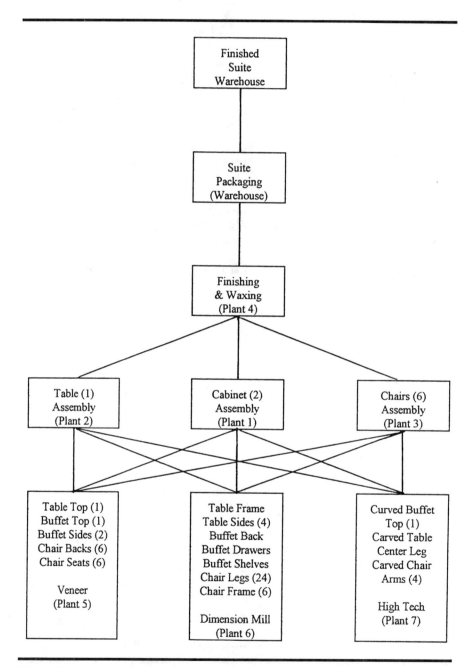

Figure 9.2   Typical Dining Room Suite Product Flow

from species to species and hack to hack. Lumber is purchased at specific times of the year, depending on the species, in order to improve appearance and manufacturability by reducing the moisture content. For example, northern pine is purchased in the summer from the Pacific Northwest in order to avoid sap in the pine. Mahogany is purchased at different times of the year, as are oak and southern pine.

Some hacks are selected for routing to the predryer building, where air is circulated to remove additional moisture. Oak is primarily routed for this additional treatment, although other species also undergo this drying operation. Moisture within the hacks is monitored by probes inserted into the lumber and data collected by computer. The air flow is adjusted by the computer as moisture is depleted from the lumber.

All hacks, including those sent to the dryer building, undergo a dry kiln baking process in a separate building. This process lasts approximately two weeks and further reduces moisture. When the moisture content is about 7%, the lumber is ready for processing. Before processing, the hacks are sent to a cooling building for several days. Approximately 250,000 board feet are dried on a weekly basis.

Plant 6, the dimension mill, is the first processing station. This plant includes all operations after drying and before machining. This includes gang rip, crosscutting, conventional ripping, mold, plane, and gluing operations. The first operation for most wood is the gang rip saw. Hacks are delivered into the Plant 6 building, broken apart, and loaded into a large machine center. The gang rip saw first cuts the boards lengthwise and makes a rough planing operation. The gang rip is a computerized operation that attempts to set the ripping blades to minimize scrap given the desired dimensions to be obtained from the lumber piece. The desired dimensions are calculated from the MRP production orders and are calculated by the shop supervisor using a Lotus spreadsheet and a personal computer.

From the gang rip saws, the lumber is transported on an automated conveyer to a series of chop cross saws. There are eight cross saws, all manually operated. The saw operators cut the lengths from the rip saw to one of six preestablished lengths. The operators are responsible for obtaining defect-free lengths, given the bill of material for the specific species being cut. The bill of material quantities are derived from the MRP production orders. The driving force for the cross-cut operators is obtaining the best yield possible from a length of lumber. After the cross-cut lumber is sent down a second conveyor, the boards are sorted by operators according to the widths obtained from the rip saws. All boards having the same length and width are placed on the same cart. At this point, roughly 50% of the lumber that was received at the rip saw is available for further production. The remaining amount, often as much as 67%, is sent to a grinder and used as fuel for the boilers.

The boards are now a standard length and width and are available for other machining operations as required on the bill of material. Depending on the item of furniture to be built, lumber is transported to Plant 1, 2, or 3 or remains in Plant 6. An additional saw cut may be required at the conventional saws, and various pieces of lumber are assembled into cores and glued. An example of a core would be several 18-inch by 1-inch by $\frac{1}{2}$-inch boards that are glued together to form a bedroom nightstand top. Another operation in the dimension mill is the planer. Several planers are available, and a core might be routed to a planer to shear the wood to an exact width and smoother surface than from the gluing operation. The dimension mill has machines that only produce a rough planed surface. At the planing operation, a defect may also be uncovered and the piece routed to another conventional saw. The operator at the conventional saw then decides how to trim out the defect and maximize the amount of good stock remaining. Also, defects can be uncovered after the gluing operation if the glue in not properly mixed or if the clamping and drying is not done properly. Any defects found as a result of the gluing operation are also routed to the conventional saws for a salvage operation.

An additional operation in the dimension mill is the molding operation. Some work pieces are selected, depending on their appearance and the bill of material, to go through one or more molding operations, sometimes called joiners. Tolerances at the molding operation are .005 to .01 inch. At this point, the color of the work piece may become important because finished pieces will be visible on the end product. There are three types of molders used at Stanley. Each type of molder has some unique capabilities as well as doing the common cuts. Parts are routed to the molder department rather than to a specific machine. The shop supervisor determines which molder is assigned to cut each work piece.

Output from the dimensional mill is: solid rectangles, stripped cores, core stock, glued rectangles, glued molded pieces, and solid molded pieces. These work pieces are then sent to the other plants at Stanleytown as required by the bill of material to make tables, chairs, and cabinets.

Plant 5 is the panel plant where veneer is produced. Vendors bring two or three samples to Stanleytown for a scheduled showing by purchasing. A sample of veneer is called a flitch. The best flitches come from the middle of the logs. When an acceptable flitch is found, the entire log is purchased. The logs have already been cut into veneers at the saw mill. About 300,000 board feet of veneer is purchased at a time. Only about 30% of the raw veneer purchased is ultimately used in furniture. The remaining 70% is scrap and is burned for fuel.

Veneer, often from the same log, or flitch, is matched by operators who create the wood patterns seen on table tops and cabinet sides. The veneer operators are more artists than craftsmen in matching patterns and textures. Operators are given six bills of material for all grades and mixtures of long, medium, and short pieces needed in order to maximize yield. As patterns emerge, cuts are made in the veneer to match the dimensions of the needed piece. At times, there may not be enough veneer left in the flitch to yield all of the needed veneer. When that occurs, the operator selects another whole flitch. The remainder of the old flitch is scrapped if no other pieces can be matched to the needed dimensions.

Finished veneer is glued to a core, joint piece, or to a solid piece, depending on the bill of material. Veneer is joined together by paper tape to create the desired patterns. The last operation is a sanding process that removes the tape once the glue has set. Veneer is also used on the edges and sides of many end items. The veneer is affixed to the cores with glue using a hot press. Some work pieces are routed to Plant 7, the high-tech area, for further operations.

Plant 7 consists of several specialized machine centers that perform unique cutting and planing operations that may be required on special work pieces. For example, a veneer core may require a precise alignment to match another veneer core. The high-technology area has a router that can align the veneers using a laser light to position the router cut. There are also diamond cutting saws available, as well as an embosser.

Plants 1, 2, and 3 have similar machining functions to produce cabinets, beds, chairs, and tables. Lumber arrives from the dimension mill and from the panel plant, where primary cutting operations are performed. Depending on the end product, various general-purpose machine centers are used to produce the needed components. Additional planers, molders, and saws are available in Plants 1, 2, and 3. The final operation in the component plants is the sanding operation. Both drum sanders and surface sanders are used. Each pass of a sander removes approximately $1/32$ inch of stock. One aspect of the final sanding operation is the scrap that is generated by sanding through a veneer. About 75% of the glued veneer scrap is caused by warpage in the work piece from moisture. Less than 10% of the final components have sanding scrap, but when it does occur the entire item must be reworked and, at times, the entire suite may be jeopardized.

The assembly operations for each plant are similar. The first operation is preassembly. As many small parts as possible are assembled before the larger work piece is sent down the assembly line. An example of a preassembled component would be a drawer for a dresser. Preassembly is not a routed operation. The individual parts are simply marked as buffer parts on the route sheets. The final assembly operation occurs on a roller conveyer. The components are joined together with nails, mortise joints, staples, or screws as required by the assembly instructions. Once assembled, the items are transported to the finishing area, designated as Plant 4.

In the finishing area, all items are air blown to remove any dust or debris. Various wood stains are prepared in the finish area, depending on the species of wood being produced. An outside company is responsible for mixing the stain according to specifications. The stain is applied to the work piece by a spray gun through a series of operators, each assigned to a specific area. The work piece is rubbed and the next coat applied. As many as six applications are applied to an end item. Tables and desk tops receive the most attention. The stain is dried in a drying booth, a lacquer applied and dried, and additional coats are applied. A final wax is applied, and the final assembly is sent to packaging.

In the packaging area, a final lacquer is applied to the table tops, hardware is placed inside drawers, and instruction sheets are added to the packaging. Cardboard containers and packing are designed for each end item. Once packed, the end item is sent to the warehouse area for consolidation with other items in the suite, if necessary, and for storage. The warehouse is a five-story building located adjacent to the factory.

Production Planning Environment

Production planning is primarily the responsibility of the corporate production scheduling department. This is a centralized function that has responsibility for the development of the master production schedule for all five Stanley factories. This department is also responsible for monitoring overall plant capacities and for the operation of the production planning and control systems. The manager of production scheduling reports to the same senior vice-president of manufacturing as do the factories' vice-presidents of production. The corporate vice-president for purchasing also reports to the senior vice-president of manufacturing. The key performance measurements for production planning are delivery performance, inventory turnover, and budget performance.

Production control is a decentralized function located in each factory's organization and reports to the factory vice-president of production. Each factory determines the need for, the number of, and the assignments of production schedulers. The production schedulers report through the line organization. Production control is responsible for execution of the production schedule and for managing short-term resource priority and capacity.

Inventory management is an important area of attention for Stanleytown factory managers. Raw material is subject to seasonality factors when different wood species must be purchased at different times of the year in order to improve quality and manufacturability. A relatively long drying period is also required for raw lumber. Current raw material inventory is about 23% of the total dollar amount. Work-in-process inventory is about 16% of the total dollar amount and has a turnover of about ten times per year. Finished goods inventory is the bulk of the total dollar amount, at about 61% of the total. More than 95% of Stanley Furniture Company's production is make-to-stock, with deliveries made from the finished warehouse stock. Additionally, one species of wood is produced at a time through the dimension mill, Plant 6, so that a cycle of products flows into the warehouse. Some amount of all types of wood species is maintained at all times in the warehouse in order to meet customer demand. An ABC classification system is used to manage the finished goods inventory. The A classification is for the items most in demand, and no stockouts are permitted to occur. The B classification is allowed to have stockouts of less than 30 days in duration. The C classification is designated for low-priority products, and stockouts are acceptable. Production of the C items occurs as the schedule permits. The key performance measurements for inventory management are the stockout level and delivery performance for the finished goods inventory. Turnover targets are established for the total inventory level. The current target calls for a turnover of about eight times per year. The Stanleytown factory is currently meeting this target.

Manufacturing lead time at the Stanleytown factory is 25 days. It is estimated that about 75% of all manufactured components require less than seven days lead time, with the longer time required for components that have machining operations performed by outside contractors. An example of an outside process is the curved wood used on an ornate cabinet. The manufacturing lead time adds the longest operation time for a batch, plus the lesser of the actual processing time or four hours for all other operations for each

part. The factory uses the manufacturing lead time for purchasing requirements and for material release. Actual manufacturing lead times are the
result of buffer management. Currently, there is a four-day buffer established
for manufactured components and an additional five days allowed for release
of raw material into the gating operation at the gang rip saws (see Figure
9.3).

Purchasing is a centralized function at the Stanley Company. The vice-
president of purchasing reports to the senior vice-president of manufacturing. Two senior purchasing managers, a buyer, and a purchasing coordinator
report to the vice-president of purchasing. One senior purchasing manager is
responsible for wood components, and the second is responsible for raw
lumber and veneer. The buyer is responsible for nonwood components. The
coordinator is responsible for overseas purchases. Purchasing has a goal of
establishing single sources for materials. A second overall goal for purchasing is developing a 15-day lead time for all items. A vendor performance
measurement system is used to monitor and improve vendor performance.

Marketing and sales is a centralized function at the Stanley Company
headquarters. The Stanley Company sells its products through independent
sales representatives to retailers. Over 3500 retailers are currently active in
the Stanley sales system. Retailers can be relatively small stores with less
than $1 million in annual sales or as large as Sears and J.C. Penney. Orders
are received from sales representatives by fax, mail, or telephone. Once an
order is received, it is entered into the sales system. Warehouse stock is
committed to the order. A credit approval process is undertaken and shipping
authorized. The traffic department arranges for trucking and load consolidation and schedules delivery to the customer location. Traffic consolidation
and shipping may require up to ten days to complete. Motor carriers usually
will not accept a load of less than 200 cubic feet. A typical trailer contains
about 2300 cubic feet. The traffic department seeks to minimize the overall
company freight charges by arranging load consolidations to several customer locations on a single full truck. The key performance measurement for
sales and service is shipping performance. Current shipping performance is
82% of all orders shipped within 30 days of receipt. The target for 1991 was
to ship 95% of all orders within 20 days of receipt.

The Stanley Furniture Company sees its competitive industry edges as
quality, delivery performance, value, and style. The company has three major
competitors: Broyhill, Thomasville, and Lexington. All three companies have
roughly the same overall market strategies. The recession in the housing
market has adversely impacted Stanley and its competitors. A small increase

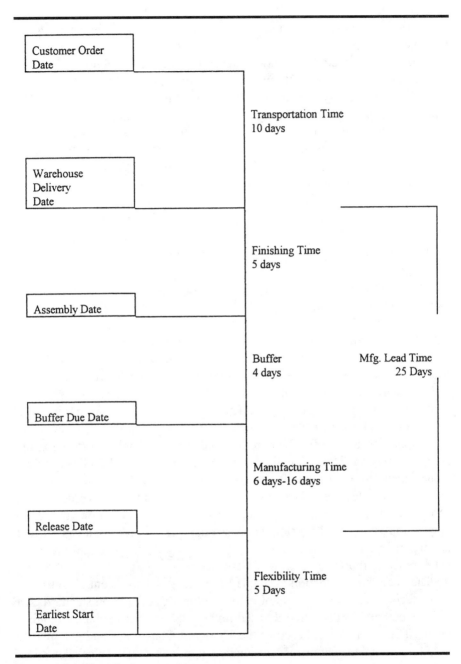

Figure 9.3 Time Scale

in overall sales has been seen in the last four months, however, as mortgage rates have fallen.

Quality assurance at Stanley is decentralized. The individual factories have inspectors who report through the line organization. At the Stanleytown factory, there are over 70 inspectors located throughout the process. The major point of emphasis is in the final assembly and finishing areas. Each hour, a finished item is selected and an internal quality audit conducted. Factory management and production supervision attend the audit session. Any defects in the item are identified and the overall quality is reviewed. The session is also devoted to correcting the problems that are found during the audit. Statistical process control charts are being implemented, although this process has only recently begun. There are no dedicated quality engineers at Stanley as the design process includes quality aspects. There is an ongoing quality education program at Stanley, and various quality projects are undertaken as the need arises.

The design process at Stanley begins with product managers. Retailers expect new products and designs each year. The average age of a design is only two years. There are four major furniture shows in the United States where Stanley and its competitors show new product designs. The product managers determine the needs of the retailers during these product shows. Product managers assign the design work to designers who create sketches. Top management, product managers, and the designers discuss the sketches and select new products and styles. The approved sketches are sent to product engineering to create a bill of material. Engineering assigns part numbers to components and the finished product and enters the bill of material in the computer system. The computer system generates master data sheets called "blue sheets" to purchasing. Purchasing adds vendors, prices, and lead times to the master data sheets. The master production schedule enters orders in the computer system to generate purchase orders. The purchase orders are reviewed to reduce the suggested purchase quantities in order to lower current factory inventory if possible. Finally, purchasing issues purchase orders to vendors for components or raw materials.

The accounting function at Stanley Furniture is also a centralized function. Accounting uses two sets of records. One set is based on standard cost roll-up methods and is primarily used for tax and security exchange requirements. The second set is based on the Theory of Constraints and is primarily used for internal management decision making. The factories are evaluated on throughput, inventory, and operating expense. The seven manufacturing plants in the Stanleytown factory are also evaluated on throughput, inventory, and operating expense. Each plant does have a budget, but the various departments do not have budgets. Plants are measured by the ratio of throughput divided by operating expense as well. Direct labor at the Stanleytown plant is about 15% of total cost. Material is about 40% of total cost, and overhead is about 45% of total cost.

Logical Structure

Based on interviews and observations, the following characteristics are present at the Stanleytown factory:

1. A relatively large number of end items are produced compared to the number of raw materials used.
2. All end items of a particular class are produced in essentially the same way.
3. There are points of divergence in the routings where components are dedicated to specific end items. This is especially true in the dimension mill, and the panel plant.
4. Some equipment is highly specialized and unique to this type of industry. This is especially true in the panel plant (Plant 5) and the dimension mill (plant 6).
5. Some equipment is general purpose. This is especially true in case goods (Plant 1), tables (Plant 2), and chairs (Plant 3).
6. A relatively large number of manufactured components are assembled into a relatively small number of end items of a particular class.
7. Components tend to be unique and specific to an end item.
8. There is considerable resource contention, especially in Plants 1, 2, 3, and 7.

These factors lead to the conclusion that the Stanleytown factory consists of two types of logical structures. The dimension mill and panel plant (Plants 5 and 6, respectively) can be described as V logical structures. The case plant, table plant, chair plant, and high-technology plant, (Plant 1, 2, 3, and 7, respectively) can best be described as A logical structures. The finishing plant (Plant 4) can best be described as an I-structure and is not be included in this study. See Figure 9.4 for a diagram of the logical structure for the dimension mill.

Additionally, characteristics indicate that the Stanleytown factory is a repetitive producer of furniture. Routings for components are fixed and established by process engineers, although supervision has some latitude in

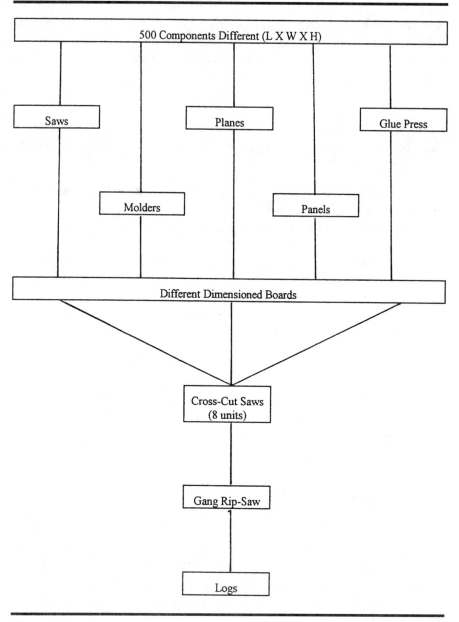

Figure 9.4   Logical Structure—Dimension Mill

assigning work to specific machine centers. Once a large run of a certain wood species begins, a relatively large number of end items are produced. Finally, production is on a make-to-stock basis rather than to specific customer orders.

PRODUCTION PLANNING AND CONTROL METHODS

Background

The Stanley Furniture Company developed an in-house MRP system for its use in 1974. An in-house system was developed because a review of existing software indicated that it did not fit the needs of the company in 1974. The system has undergone modification over time and until 1990 was used for master production scheduling, capacity requirements planning, MRP calculations, shop floor scheduling, and purchasing. In 1990, the company adopted the Theory of Constraints production system at Stanleytown. Primary use is made of the drum-buffer-rope technique, the five-step focusing process, and the throughput, inventory, and operating expense performance measurement system. Training began in May 1990 and continued into early 1991. All management and floor supervisors participated in a one-day training session held by an outside consultant. A new position was created for a MBA educated in the Theory of Constraints philosophy. Additionally, four internal experts, called "Jonahs," have received education in the Theory of Constraints. Additional education is planned for the internal Jonahs.

Currently, the MRP system is used for master production scheduling and for purchased parts scheduling. Also, the bill of material and routing files are maintained. Additionally, some Just-in-Time activities have been implemented, along with the Theory of Constraints scheduling system. The Stanleytown factory is organized along the factory-within-a-factory concept, and there is an active program to reduce setups and run sizes throughout the plant. Emphasis is on a total quality management program and supplier involvement in the design process. However, no use is made of the kanban method, homogeneous master production scheduling, or leveled master schedules. The setup reduction program has recently been refocused on the machine centers that are believed to be near constraints throughout the various plants. Management at the Stanleytown factory believes the bills of material used in the production planning system are very accurate and pose no real problems in the execution of the production plans. Changes to the bills of material are infrequent and are tied into new product implementation. Prior to the introduction of new products, a pilot build is conducted to assure proper component fit. During the pilot build, any problems with the bill of material are identified and corrective action taken.

Routings at the Stanleytown factory do present problems to the production planning and control system at times. Management estimates that the majority of the production routings are at least 50% accurate, but only about 20% of the routings are 100% accurate. There is no audit function for identifying erroneous routings. Engineering changes to the routings are infrequent. Management appears to rely on the skills of the employees and supervisors to know how to produce a component rather than on the actual routing sheets. Routing sheets appear to be used primarily for transferring a load of material between departments and for load identification.

The order policy used in the MRP planning function is lot-for-lot, with a minimum order quantity of 150 units. Manufacturing lead time in the MRP system is 25 workdays. Total quoted customer lead time is 45 workdays, which includes the 25 days plus 20 days for distribution and administration. The lead times are used only for purchasing information. No safety stock is used in the system. The MRP system uses weekly buckets to supply requirements to the purchasing system.

Master Production Scheduling

The development of the master production schedule (MPS) is centralized at the Stanley Company headquarters. An MPS is manually prepared for each of the five factories based on the sales forecast. The annual sales forecast is prepared in the fall for the next year. Each month a revision is prepared for a rolling four-month time period. The forecast uses a monthly time bucket and aggregates the product lines in terms of whole suites of furniture: dining room suites, bedroom suites, entertainment units, home office units, youth suites, and occasional units. The forecast is approved by upper management and is the basis for financial plans as well as establishing sales goals.

The MPS uses the forecast as a building block. In the production planning phase, adjustments are made to the monthly forecast based on the known production capabilities at each factory and any desired inventory adjustment. The end result is the MPS in weeks for each factory for the next 4.5 months. The first page is the MPS for Stanleytown Plants 1, 2, and 3 and the MPS for Waynesboro, identified as "Northern." Page two is the MPS for Lexington, Robbinville, and West End, identified as "Southern." The first 45 days of the MPS are frozen, although changes to the MPS average about one per week. These changes are usually the result of internal changes such as scrap over the planned amount. The most recent sales history for the forecasted suite is used by the master production scheduler to determine an expected stockout date for each item (SKU) in the warehouse based on the ABC classification scheme previously discussed. As a result of this calculation, a priority schedule is developed and a cutting list is prepared. The cutting list identifies the wood species and linear feet to be cut at the gating operation. The MPS is approved at a monthly meeting attended by top management. The 45 days used in the MPS as a frozen time period permits cuttings to be made at the gating operation from dried lumber in sufficient quantities to build the desired amount of furniture.

Once approved, the MPS is loaded into the MRP system to generate purchasing requirements. A final assembly schedule (FAS) is prepared from the MPS. The FAS is prepared for the chair plant, cabinet plant, and table plant. The scheduled time on the final assembly schedule is based on the engineering run rates plus a 45-minute setup time between the runs of different suites. The FAS is the drum used to establish the pace of production for the Stanleytown factory. Because management believes that the marketplace is the current constraint to the system, the final assembly lines are used to establish the internal constraints by adjusting work force assignments. In this way, the FAS is kept equal to the MPS.

Production is monitored on a daily basis, and adjustments to capacity are made using overtime to keep the assembly lines on schedule. The master production scheduling department maintains records of actual production and makes adjustments as required to support future schedules by recommending work force adjustments.

Priority Planning

Part priorities are established by two general approaches. For purchased components, MRP logic is used, allowing a 45-day lead time for all parts. The MPS identifies the due date for the assembly of the end items in weekly time periods. The production requirements for end items exploded, and time phased via the MRP calculation and purchase orders are issued. Management estimates that about 75% of all purchased items can be delivered within seven production days.

Part shortages are identified on a report called "The Scheduled Production Shippable Deficiencies Aged." This report indicates the part shortage quantity, the suite number, the item number, the promise date, and the number of days the part is deficient from assembly. Three expediters, who report to a manager of purchasing, monitor this report and take necessary actions. Typical items on this report would include mirrors, glass table tops, and parts routed to an off-site contractor for a special cutting operation.

Part priorities for manufactured parts are scheduled according to a four-day buffer established prior to the assembly lines, at the drum. Material is released into the gating operation according to the FAS date, adjusted for lead time. The lead time is calculated for each manufactured component going into the final assembly. The lead time is calculated by adding the longest operation time in the part's routing for processing the entire batch and adding four hours for every other operation or the actual processing time if it is greater than four hours (see Figure 9.5). This calculation determines the start dates for all components for final assembly. The rough mill (Plant 6) and the panel plant (Plant 5) are permitted to start a part ten days prior to the calculated start date. The due dates are communicated to the shop on a buffer schedule report. The buffer schedule report has the following categories: scheduled quantity, cutting number, suite number, component number, part description, buffer due date, assembly date, buffer zone, start date, and part location. The buffer zone is a designation from zero to four and indicates the position of the part order quantity with respect to the assembly date. The entire batch quantity is scheduled to arrive on the buffer due date. However, the assembly is scheduled over many days so that not all components in the batch are needed in the first buffer zone on the scheduled due date. The floor supervisors and buffer manager know this situation and adjust priorities accordingly. Actual shortages in buffer region 1 were 32 parts over a two-week time period.

The buffer is constantly monitored by a buffer manager, who takes necessary actions to ensure the parts are expedited to the assembly lines. A Buffer Hole Check Sheet has been developed to identify the causes of parts missing from the buffer. Causes of missing parts are investigated and changes made as required. The Buffer Hole Check Sheet has only recently been established. In addition to the Buffer Hole Check Sheet, a priority list is developed. This report identifies parts where several batches of different components going to the same final assembly are scheduled to arrive at the buffer at the same time. Part priorities are established by identifying the start dates for the batches. The earliest start date is produced first.

Because the drum has been established at the assembly line, all manufactured components are scheduled across the constraint. As a result, all manufactured parts are identified on the buffer schedule.

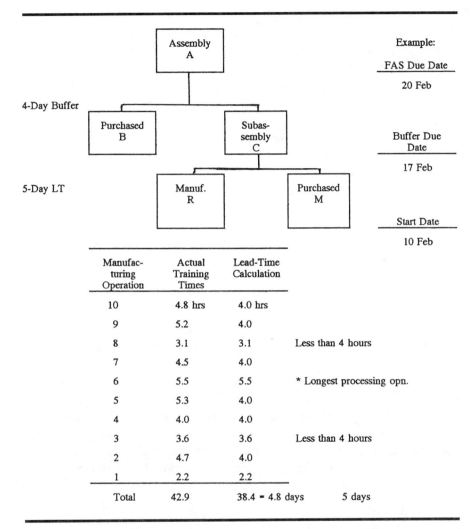

Manufac- turing Operation	Actual Training Times	Lead-Time Calculation	
10	4.8 hrs	4.0 hrs	
9	5.2	4.0	
8	3.1	3.1	Less than 4 hours
7	4.5	4.0	
6	5.5	5.5	* Longest processing opn.
5	5.3	4.0	
4	4.0	4.0	
3	3.6	3.6	Less than 4 hours
2	4.7	4.0	
1	2.2	2.2	
Total	42.9	38.4 = 4.8 days	5 days

Figure 9.5   Lead Time Calculation

Capacity Planning

Capacity planning was originally accomplished by using computer reports generated from the MRP system. The machine load was determined for each machine center using the standard hours from the engineering routing file and the MRP part requirements. Capacity is currently viewed in terms of standard labor hours.

Long-term capacity planning is accomplished on an annual basis as part of the development of the annual sales forecast. One-year and three-year plans are developed and reviewed by top management. Equipment acquisition decisions and work force levels are determined using the long-term capacity planning output from the MRP system. A comparison is also made between the previous year's forecast and the work force used to produce the actual output.

During the year, capacity is monitored as part of the monthly master production scheduling process. The MPS is used to determine the work force levels by comparing the expected production with the previous period's production. The capacity schedule and MPS are kept synchronized as part of the scheduling process. The primary machine centers that are monitored during the master production scheduling process are the assembly lines. All other machine centers are compared against current work force level required to support the current assembly line rate of production. Management does not treat any machine center other than the assembly lines as a constraint to the system.

Adjustments to short-run capacity are made by adjusting the hours worked by the current work force. If actual requirements are greater than the available capacity, overtime is scheduled. If actual requirements are less than the available capacity, a shortened work week may be scheduled. If the available capacity exceeds requirements for a very short time or is isolated in a small number of the machine centers for a short amount of time, material may be released early into the shop. The work force is not formally measured on efficiency comparisons to standards, but rather on delivery to the buffer due dates. However, there is a management concern that employees not be idle for any length of time, and supervisors continue to informally monitor efficiency.

Priority Control

Priority control is focused at the buffer prior to the assembly lines. A buffer manager is assigned responsibility for taking all actions necessary to ensure parts arrive at the buffer in time to support the final assembly line schedule. The buffer manager uses informal expediting throughout the factory to maintain a flow of parts into the assembly area. No daily dispatch list used at the Stanleytown factory, nor is a shortage list prepared. The buffer report schedule is the source for expediting parts as the need arises in the production process.

The only other priority control function occurs in the finished goods warehouse. Production is make-to-stock; therefore, the finished goods inventory decouples the customer orders from the actual production activities. The traffic department develops truckload consolidations based on shipment weights and geographic considerations in order to minimize transportation costs. Shipping orders are sent from the traffic department to the warehouse each day. A packet consisting of shipping labels, a copy of the invoice, and a sheet providing the customer name and address is prepared for each shipment. A pick list is prepared by the warehouse supervisor for each floor in the five-story warehouse. A chalkboard is used to identify the loads to be shipped during the day and the status of the loads. As the picks are executed and the loads built in a layout area, a bill of lading is prepared. The loads consisting of the customer orders are then placed in the proper trucks for shipping. A daily report is written by the warehouse supervisor to the traffic department identifying the shipments made during the day.

Capacity Control

Capacity control is a decentralized activity at each of the factories in the company. Under the Theory of Constraints, the Stanleytown factory does not formally measure equipment utilization or labor efficiency. The work centers are managed informally by the line organization to support the buffer schedule rather than to maintain work. Line supervision reviews the current work force levels each month, but this is an informal process in which throughput is compared to the amount scheduled to support the buffer. If actual capacity available exceeds requirements, the workers are assigned to nonproduction activities. If capacity available is less than the requirements, overtime is requested. The overtime request is reviewed and approved by the manager of the plant involved.

Capacity at the assembly lines is controlled by the manufacturing organization. The output from each assembly line is monitored on a daily basis against the FAS. Any arrearages can usually be made up in the normal assembly process by the work force adjusting its rate of production. If the arrearage increases to over four hours of work, overtime is scheduled. As part of the MPS development, the work force level on the assembly lines is monitored and adjustments made as required. As the assembly lines are monitored as part of the MPS development, adjustments to the work force level are seldom made inside the four-month planning horizon.

KEY PROBLEM AREAS BY PRODUCTION FUNCTION

Master Production Scheduling

The first area of concern in the development of the MPS is the seasonality in the demand for the final product. The months with the lowest sales volume are December, January, and February. The months with the highest sales volume are May, June, and July. The seasonality complicates the development of the MPS because of management's desire to maintain a stable work force in the factories. The second area of concern is the seasonality and lead time required for raw lumber from the lumber mills. Various different wood species must be purchased at different times of the year to reduce the moisture content of the logs and improve the appearance and manufacturability of the lumber. This seasonality differs among the wood species. The raw lumber seasonality does not have any relationship to final customer demand seasonality. As a result of the raw lumber seasonality, a significant amount of raw lumber is on-site at the factories. The third area of concern is the cycle time used in the MPS for the production of furniture from the different species. The production cycle time for finished goods replenishment is about 65 days but does not coincide with the final customer demand seasonality.

A fourth concern is the relationship of the FAS to the MPS. In Theory of Constraints, the flow of production for the entire factory is established by the drum. The drum is the FAS for the assembly lines. The FAS must be kept synchronized with the MPS. There is concern that another workstation is or may become the actual constraint as demand increases. If a new constraint emerges, the proper use of the FAS as the drum is not clear.

Priority Planning

The major area of concern for priority planning is the yields from the raw lumber at the gating operation and at the panel plant. Yields from raw lumber may be as low as 40% in terms of the square foot of the original log. Yields are difficult to predict, and small variations may have significant negative consequences, especially at the end of a run of a type of wood species. A second area of concern is the control over the gating operation at the dimension mill. The dimension mill is permitted to start lumber up to ten days before the calculated start date to support the buffer. Additionally, the output in dimensioned pieces may be different than the amount required from the bills of material. A third concern in priority planning is the difficulty experienced by management in identifying the system's constraint. The factory is using the final assembly lines as the constraints, but there is a belief that manning at other machine centers may be the constraints, especially as furniture demand rises. A fourth area of concern in priority planning is a general feeling among some manufacturing supervision that the lack of interim start dates and due dates causes discomfort. Supervisors are sometimes not sure of their position with respect to the buffer.

Capacity Planning

The key concern in capacity planning is difficulty in calculating the required work force levels at nonconstraint machine centers. Data used for the standard hour calculation are regarded as inaccurate. A second concern is the coordination necessary concerning capacity planning and the development of the MPS. The master production scheduling process assumes capacity exists throughout the factory to support the expected production level. Only the assembly lines are reviewed for the work force levels when the MPS is determined. A final concern in capacity planning is the former performance measurement system's influence on current employee behavior. While the factory no longer formally calculates machine utilization or work force efficiencies, management is uncomfortable paying workers to be idle for any period of time.

Priority Control

The major concern in priority control is the management of the buffer at the assembly line. About 10% of all items scheduled to be in the buffer are not available and must be expedited. A buffer manager is assigned to take all necessary actions to expedite components into the assembly area, and the factory has implemented a buffer hole investigation report. However, components are continuing to be expedited into the assembly lines. A second concern is the absence of some parts from the buffer schedule report itself. Components that are needed in the preassembly area for prebuilds are not in the bill of material and, therefore, are not scheduled to arrive at the proper time into the buffer. A third concern is the expediting that is occasionally required for purchased components as a result of errors in the purchasing system or late supplier deliveries.

Capacity Control

The major area of concern in capacity control is the belief that protective capacity needs to be established throughout the factory. A second area of concern is the overtime that may be required at times for machine centers other than the assembly lines. Expediting from the buffer is sometimes viewed as a surprise to manufacturing, and short-term overtime is used to supply the necessary components.

WITHIN-CASE ANALYSIS OF STANLEY FURNITURE COMPANY PLANTS 5 AND 6

Data gathered from the previsit survey and on-site interviews identified several areas of concern for Stanley Furniture Company management in terms of the development and management of the production planning and control system. Two logical structures exist in the Stanleytown factory. Plant 5 (the dimension mill) and Plant 6 (the panel plant) are V-structures. Plant 1 (case goods), Plant 2 (tables), Plant 3 (chairs), and Plant 7 (high technology) are best described as A logical structures and are analyzed in the within-case analysis that follows this case analysis. Plant 4 (the finishing plant) is an I-structure and is excluded from the study.

Table 9.1 summarizes the methods used to plan and control production in Plants 5 and 6. Table 9.2 summarizes the key problems encountered by management in planning and controlling production in Plants 5 and 6. Table 9.3 summarizes the critical characteristics and impacts on the production planning and control system in the V-structures of Plants 5 and 6.

CRITICAL CHARACTERISTICS AND IMPACTS BY FUNCTION FOR THE V-STRUCTURE PLANTS

Master Production Scheduling

The first critical characteristic that impacts the MPS function for the V-structure plants at Stanleytown is the centralization of the MPS function. The second characteristic is the high product seasonality for furniture. The third characteristic is the 30-day customer lead time for furniture made from all species of wood and styles of furniture. The centralization of the MPS function, the product seasonality, and the 30-day customer delivery lead time

Table 9.1   Production Planning and Control System Methods Used at Stanley Furniture Company, Stanleytown, Virginia

V-STRUCTURE PLANTS

Master Production Scheduling
1. The assembly requirements are exploded by the MRP system from the MPS system on the weekend regeneration to create requirements for the dimension mill and the panel plant.
2. A cutting schedule is developed from the MPS based on production orders, availability of dried species of wood, and the log purchase planning cycles.
3. The changes in the buffer status are determined each day and adjustments are made in the buffer status report.

Priority Planning
1. The MRP system uses lead time to offset requirements for raw lumber to the gating operation at the dimension mill and the panel plant.
2. The buffer due dates are determined for components that feed the constraint (the assembly lines) and are communicated back into the shop by the buffer schedule.
3. The buffer manager monitors the buffer status on a continuous basis and communicates changes and requirements to the manufacturing supervisors in the dimension mill and panel plant.
4. Manufacturing supervision adjusts production to meet the requirements indicated by the buffer manager.

Capacity Planning
1. Capacity for work centers, other than the assembly lines, is determined by each manufacturing supervisor based on the existing work force level and the production rates to be used on the assembly lines.
2. Work force adjustments are made by reassigning existing employees, cross-training, and hiring new employees.

Priority Control
1. Priority control occurs as requirements change on the buffer schedule or as communicated by the buffer manager to floor supervision.
2. Expediting occurs as required to support the buffer.

Capacity Control
1. Overtime is used to support the buffer requirements if required.
2. The gating operations at the dimension mill and in the panel plant are permitted to launch work up to ten days prior to the calculated start date in order to increase manufacturing efficiencies throughout the factory.

Table 9.2   Summary of Key Problem Areas at Stanley Furniture Company, Stanleytown, Virginia

---

V-STRUCTURE PLANTS

Master Production Scheduling
1. Master scheduling across plants that have horizontal and vertical dependencies creates difficulties in coordinating each plant's FAS with the MPS.
2. Yields from raw logs are variable and difficult to predict.

Priority Planning
1. There is difficulty in planning material release at the dimension mill and in the panel plant because of the unpredictability of the yields.
2. The internal system constraint is difficult to determine because of the lot sizing and processing batch policy constraints.
3. Some manufacturing supervisors feel that they do not know their schedule position because no intermediate due dates are used at nonconstraint operations, which adversely impacts work force assignments.

Capacity Planning
1. A lack of information exists about the work force required to support a certain level of production dictated by the assembly line rates.

Priority Control
1. Components missing from the first buffer zone cause expediting by the buffer manager while parts from these same work centers are already in the buffer.
2. There is a feeling among some manufacturing supervisors that they do not know their schedule position, which causes some misallocation of resources.
3. The former performance measurement system for monitoring machine utilization and labor efficiencies still influences some manufacturing operations, causing a misallocation of resources.

Capacity Control
1. There is a feeling that the Stanleytown factory needs to add protective capacity throughout the factory to reduce temporary bottlenecks.
2. Overtime to support the buffer is unpredictable and creates disruptions in the feeding operations.

---

combine to cause the forecast to be used to generate production requirements for both the panel plant and dimension mill rather than actual customer orders. The centralization of the MPS function, the high seasonality of end products, and the delivery time combine with the use of forecasts to cause the production requirements to be relatively unstable outside the MPS planning horizon.

The fourth characteristic is vertical and horizontal dependencies for the panel plant and dimension mill parts that are required to support the chair plant, cabinet plant, high-technology plant, and table plant. The fifth characteristic is the relatively long cycle times for manufacturing different species

Table 9.3   Critical Characteristics of V-Structures at Stanley Furniture Company, Stanleytown, Virginia

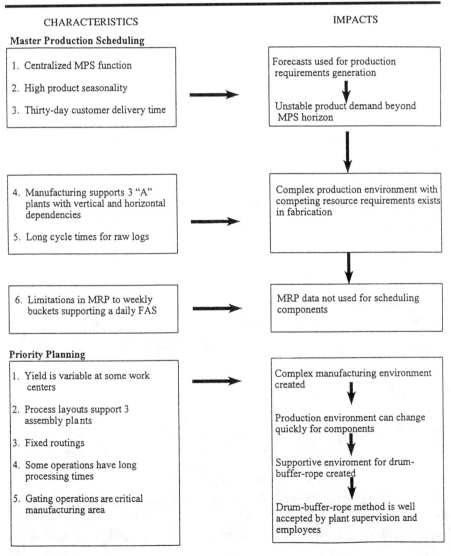

CHARACTERISTICS

Master Production Scheduling

1. Centralized MPS function

2. High product seasonality

3. Thirty-day customer delivery time

4. Manufacturing supports 3 "A" plants with vertical and horizontal dependencies

5. Long cycle times for raw logs

6. Limitations in MRP to weekly buckets supporting a daily FAS

Priority Planning

1. Yield is variable at some work centers

2. Process layouts support 3 assembly plants

3. Fixed routings

4. Some operations have long processing times

5. Gating operations are critical manufacturing area

IMPACTS

Forecasts used for production requirements generation

Unstable product demand beyond MPS horizon

Complex production environment with competing resource requirements exists in fabrication

MRP data not used for scheduling components

Complex manufacturing environment created

Production environment can change quickly for components

Supportive enviroment for drum-buffer-rope created

Drum-buffer-rope method is well accepted by plant supervision and employees

Table 9.3   Critical Characteristics of V-Structures at Stanley Furniture Company, Stanleytown, Virginia (continued)

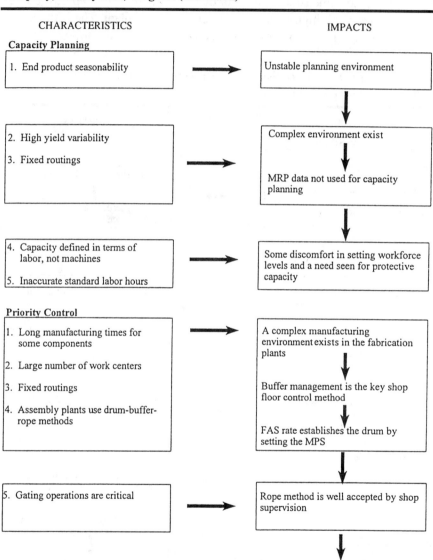

of woods through the dimension mill and panel plant. The long cycle times are the result of the same species of wood being dried in the dryer building prior to being ready to be processed in the dimension mill and panel plant.

Table 9.3   Critical Characteristics of V-Structures at Stanley Furniture Company, Stanleytown, Virginia (continued)

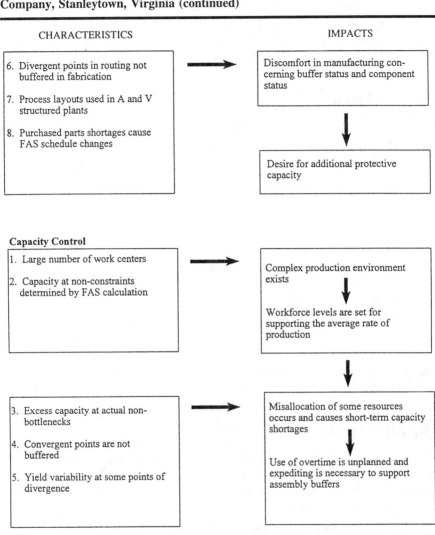

CHARACTERISTICS	IMPACTS
6. Divergent points in routing not buffered in fabrication	Discomfort in manufacturing concerning buffer status and component status
7. Process layouts used in A and V structured plants	
8. Purchased parts shortages cause FAS schedule changes	Desire for additional protective capacity

Capacity Control

1. Large number of work centers	Complex production environment exists
2. Capacity at non-constraints determined by FAS calculation	Workforce levels are set for supporting the average rate of production
3. Excess capacity at actual non-bottlenecks	Misallocation of some resources occurs and causes short-term capacity shortages
4. Convergent points are not buffered	Use of overtime is unplanned and expediting is necessary to support assembly buffers
5. Yield variability at some points of divergence	

The interplant dependencies and the long cycle times between species combine with the unstable product demand beyond the planning horizon to cause the V-structure plants to operate in a relatively complex production environment with competing resource requirements.

The sixth characteristic is the limitation of the MRP system to weekly buckets even though production of final suites occurs daily in the A-structure plants. The limitation of the MRP system and the complex production environment combine to keep MRP data from being used to master schedule the V-structure plants.

Priority Planning

The first characteristic is the yield variability in a number of work centers in the V-structure plants. The second characteristic is the dependencies caused by the process layout used in the A-structure plants. The third characteristic is the use of fixed routings in the V-plants. The fourth characteristic is the relatively long manufacturing times required for some products. The fifth characteristic is the critical nature of the gating operation to the V-structure. All of the characteristics combine to cause the V-plants to operate in a relatively complex production environment. The complex production environment that exists in the V-plants is subjected to rapid changes due to yield variability, requirements changes, machine downtime, and other factors common in a complex manufacturing environment. As a result of the fluctuations in the production environment, an environment that supports the application of drum-buffer-rope to plan priorities is created. As a result of the production environment requiring the use of the drum-buffer-rope method, the plants' management and employees readily support the drum-buffer-rope application.

Capacity Planning

The first characteristic that impacts the capacity planning function in the V-plants is the high seasonality in end item furniture suites. The seasonality causes an unstable capacity planning environment that requires leveling to support a stable work force.

The second characteristic is the yield variability in a number of work centers in the V-structure plants. The third characteristic is the use of fixed routings in the V-plants. The yield variability and the use of fixed routings cause a complex production environment to exist in the V plants. The unstable planning environment and the complexity of the production environment combine to cause MRP data to be insufficient for planning capacity in the V-structure plants.

The fourth characteristic is defining capacity in terms of labor rather than in terms of equipment. The fifth characteristic is the relatively inaccurate standard labor times in the routing file, which inhibits capacity calculations. The inability to use MRP data combined with the use of labor to define capacity and the inaccurate standard hour data cause management to feel uncomfortable in setting work force levels in the V-plants and create a feeling that there is a need to increase the overall levels of protective capacity in the V-structures.

Priority Control

The first critical characteristic that impacts the priority control function in the V-structure plants is the long total processing time required for some parts. The second characteristic is the large number of work centers in the plants. The third characteristic is the use of fixed routings. The fourth characteristic is the use of the drum-buffer-rope method in the A-structure plants. The long manufacturing times, the large number of work centers, the use of drum-buffer-rope in the A-structure plants, and the use of fixed routings combine to create a complex manufacturing environment that supports the use of buffer management as the key shop floor control methods in the V-structure plants. The use of buffer management causes the linking of the V-structure plants with the assembly lines in the A-structure plants by using the FAS as the drum that ties the V-plants' gating operations with the rope.

The fourth characteristic is the criticality of the gating operations in the Stanleytown factory. The gating operations are successfully managed by the use of the rope component of the drum-buffer-rope method for controlling priorities. As a result of the success of the rope component and the acceptance of the drum-buffer-rope methods in the A-structure plants, the use of the drum-buffer-rope method is well accepted in the V-structure plants.

The fifth characteristic is the existence of divergent points in the V-plants' routing that are not buffered. The sixth characteristic is the process layouts used in the plants. The seventh characteristic is the changes in the FAS that are caused by late deliveries of purchased parts. The lack of buffers, the use of process layouts, and the FAS changes combine to cause some discomfort in the manufacturing areas concerning the component parts' buffer status. This discomfort results in a desire to add protective capacity in the V-plants.

Capacity Control

The first critical characteristic that impacts the capacity control function in the V-structure plants is the relatively large number of work centers in the V-plants. The second characteristic is the use of the FAS to determine the work force requirements at nonconstraint operations. The number of work centers and the use of the FAS to calculate capacity at nonconstraints combine to create a relatively complex capacity planning environment in the V-plants. The complex environment causes the work force levels to be set for the average rate of production rather than the actual work force required.

The third characteristic is the excess capacity at the nonbottleneck operations and the desire by shop floor supervision to make sure that employees are not idle. The fourth characteristic is the lack of buffers at divergent points in the V-plants. The fifth characteristic is the amount of yield variability at points of divergence. As a result of the excess capacity at nonconstraint operations, the lack of buffers at divergent points, and the yield variability at points of divergence, there are misallocations of some resources and materials in the V-sructure plants, which causes short-term capacity shortages. As a result of the short-term capacity shortages, unplanned overtime and some expediting are necessary to support the assembly buffers.

WITHIN-CASE ANALYSIS OF STANLEY FURNITURE COMPANY A-STRUCTURE PLANTS 1, 2, 3, AND 7

Data gathered from the previsit survey and on-site interviews identified several areas of concern for Stanley Furniture Company management in terms of the development and management of the production planning and control system. Two logical structures exist in the Stanleytown factory. Plant 1 (case goods), Plant 2 (tables), Plant 3 (chairs), and Plant 7 (high-technology items) are A-structures. Plant 5 (the dimension mill) and Plant 6 (the panel plant) are V-structures. Table 9.4 summarizes the methods used to plan and control production in the A-structure plants. Table 9.5 summarizes the key problem areas that were identified in the A-structure plants. Table 9.6 identifies the critical characteristics and the impacts on the production planning and control system in the A-structure plants.

Table 9.4   Production Planning and Control System Methods Used at Stanley Furniture Company, Stanleytown, Virginia

A-STRUCTURE PLANTS

Master Production Scheduling
1. The sales forecast is developed for the next year in the fall and updated monthly.
2. The master production scheduler uses the sales forecast, current warehouse inventory level, and recent sales activity to develop the schedule, with the first 4.5 months frozen.
3. As shipments are made from the warehouses to fill customer demand, the inventory levels are monitored. If the inventory falls below a predetermined level, the suite number appears on a projected stockout list. The suite is assigned to a time period in the MPS where replenishment can occur prior to the stockout.
4. Capacity at each of the five Stanley factories is reviewed and adjustments are made to the MPS, if required, to assign production to each factory.
5. The FAS is developed for the Stanleytown assembly lines in daily buckets for the first week based on the MPS.
6. The FASs are monitored on a daily basis and overtime scheduled as needed so that no adjustment is made to the MPS.

Priority Planning
1. The FASs determine the assembly line priorities.
2. The warehouse receives a daily shipping report from the traffic department that establishes a ship schedule which meets customer order due dates.
3. The MRP system uses MPS requirements, the bills of material, and routings to determine purchased part requirements.
4. A buffer management report is developed to establish a buffer due date for manufactured parts.
5. A priority list report establishes priorities for manufactured parts if more than one part is scheduled for the same buffer due date.
6. A buffer manager adjusts priorities as required based on holes in the buffer.

Capacity Planning
1. Capacity is planned in conjunction with the development of the MPS on a monthly basis.
2. The Stanleytown assembly lines determine the plants' capacities.
3. Work force levels are adjusted in the fabrication areas as required by the MPS.

Priority Control
1. The FAS is executed with about one change per week required.
2. Priority control occurs by management of the buffer as the buffer manager communicates requirements to the manufacturing organization. About 15 parts per week are missing from the first buffer region and are expedited in each plant.
3. Expediting occurs as required to support the buffer and the FAS.

Table 9.4   Production Planning and Control System Methods Used at Stanley Furniture Company, Stanleytown, Virginia (continued)

Capacity Control
1. Capacity is controlled by using overtime on the assembly lines and in fabrication areas to maintain the FAS requirements.
2. Employees are assigned to nonproduction tasks as required rather than pulling work into the assembly lines.

Table 9.5   Summary of Key Problem Areas at Stanley Furniture, Stanleytown, Virginia

A-STRUCTURE PLANTS

Master Production Scheduling
1. Seasonality in customer demand for Stanley products must be reflected in the MPS.
2. Seasonality and long lead times for raw material must be reflected in the MPS.
3. The existence of an actual constraint is difficult to determine in the fabrication areas.
4. The cycle time for wood species in production is difficult to match with actual customer demand, resulting in adjustments in the MPS in order to manage inventory levels.
5. Customer orders trigger stock replenishment orders from the warehouse that must be inserted into the MPS. Minimum order quantities for replenishment orders must be considered in developing the MPS.

Priority Planning
1. Yields from operations are variable and difficult to forecast.
2. The internal system constraint is difficult to determine.
3. Some items do not appear on the buffer schedule report.
4. There is a feeling among some manufacturing supervisors that they do not know their schedule position because no intermediate due dates are used at nonconstraint operations.

Capacity Planning
1. A lack of information exists on the work force level needed to support the MPS at subassembly areas.

Priority Control
1. Components are missing from the buffer in the first zones, causing expediting by the buffer manager.
2. Some purchased components must be expedited from outside contractors.
3. Existing policy constraints concerning lot sizing and transfer batches inhibit flow and cause temporary constraints.

Capacity Control
1. Overtime is unpredictable in some operations that support the assembly line.

Table 9.6　Critical Characteristics of A-Structures at Stanley Furniture Company, Stanleytown, Virginia

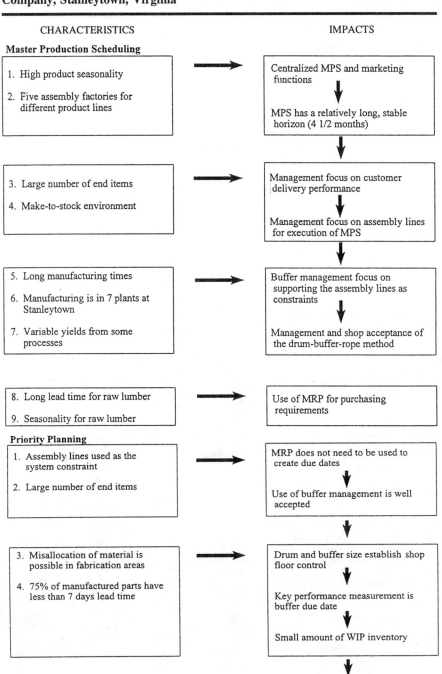

CHARACTERISTICS　　　　　　　　　　　　　　IMPACTS

Master Production Scheduling

1. High product seasonality

2. Five assembly factories for different product lines

→ Centralized MPS and marketing functions

↓

MPS has a relatively long, stable horizon (4 1/2 months)

↓

3. Large number of end items

4. Make-to-stock environment

→ Management focus on customer delivery performance

↓

Management focus on assembly lines for execution of MPS

↓

5. Long manufacturing times

6. Manufacturing is in 7 plants at Stanleytown

7. Variable yields from some processes

→ Buffer management focus on supporting the assembly lines as constraints

↓

Management and shop acceptance of the drum-buffer-rope method

↓

8. Long lead time for raw lumber

9. Seasonality for raw lumber

→ Use of MRP for purchasing requirements

Priority Planning

1. Assembly lines used as the system constraint

2. Large number of end items

→ MRP does not need to be used to create due dates

↓

Use of buffer management is well accepted

↓

3. Misallocation of material is possible in fabrication areas

4. 75% of manufactured parts have less than 7 days lead time

→ Drum and buffer size establish shop floor control

↓

Key performance measurement is buffer due date

↓

Small amount of WIP inventory

↓

Table 9.6   Critical Characteristics of A-Structures at Stanley Furniture Company, Stanleytown, Virginia (continued)

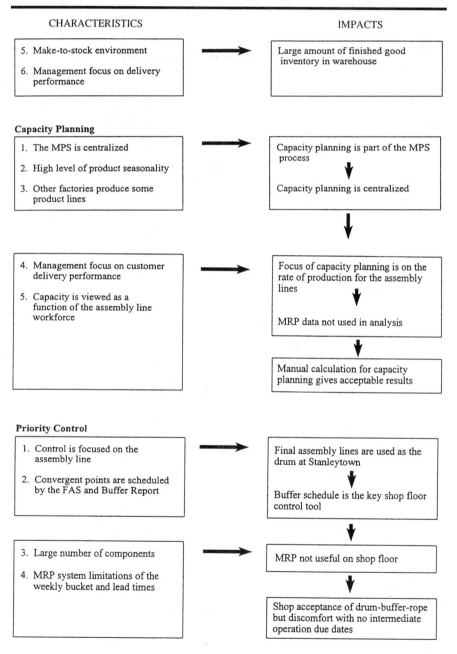

CHARACTERISTICS   IMPACTS

5. Make-to-stock environment

6. Management focus on delivery
   performance

→ Large amount of finished good
   inventory in warehouse

Capacity Planning

1. The MPS is centralized

2. High level of product seasonality

3. Other factories produce some
   product lines

→ Capacity planning is part of the MPS
   process

   Capacity planning is centralized

4. Management focus on customer
   delivery performance

5. Capacity is viewed as a
   function of the assembly line
   workforce

→ Focus of capacity planning is on the
   rate of production for the assembly
   lines

   MRP data not used in analysis

   Manual calculation for capacity
   planning gives acceptable results

Priority Control

1. Control is focused on the
   assembly line

2. Convergent points are scheduled
   by the FAS and Buffer Report

→ Final assembly lines are used as the
   drum at Stanleytown

   Buffer schedule is the key shop floor
   control tool

3. Large number of components

4. MRP system limitations of the
   weekly bucket and lead times

→ MRP not useful on shop floor

   Shop acceptance of drum-buffer-rope
   but discomfort with no intermediate
   operation due dates

Table 9.6 Critical Characteristics of A-Structures at Stanley Furniture Company, Stanleytown, Virginia (continued)

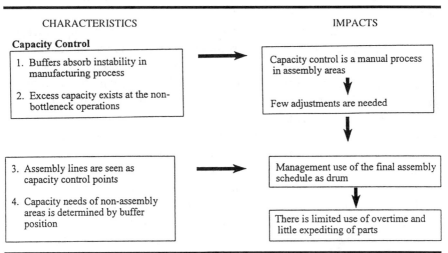

CHARACTERISTICS   IMPACTS

Capacity Control

1. Buffers absorb instability in manufacturing process

2. Excess capacity exists at the non-bottleneck operations

Capacity control is a manual process in assembly areas

Few adjustments are needed

3. Assembly lines are seen as capacity control points

4. Capacity needs of non-assembly areas is determined by buffer position

Management use of the final assembly schedule as drum

There is limited use of overtime and little expediting of parts

CRITICAL CHARACTERISTICS AND IMPACTS BY FUNCTION FOR THE A-STRUCTURE PLANTS

Master Production Scheduling

The first critical characteristic that impacts the master production scheduling function in the A-structure plants at Stanley Furniture is the high product seasonality for furniture. The second key characteristic is the use of five different factories in the Stanley Company to produce different types of furniture. The product seasonality and the need to support five different factory production capabilities causes the Stanley Company to centralize the marketing and MPS functions. The centralization of the marketing and multiplant MPS functions (horizontal and vertical) causes the company to use a relatively long (4.5 months) stable planning horizon rather than a shorter, more responsive period.

The third characteristic that impacts the MPS function is the large number of end items that are produced by the Stanleytown factory. The fourth characteristic is the existence of a make-to-stock product environment. The large number of end items and the make-to-stock environment combine with the relatively long stable planning horizon used in the MPS to cause management to focus on delivery performance as the key performance measurement for the factory. As a result of the criticality of delivery performance, management focuses on the production rates on the assembly lines at the Stanleytown factory to execute the MPS.

The fifth characteristic is the relatively long manufacturing times required for some components. The sixth characteristic is the horizontal and vertical dependencies of the seven manufacturing plants used at Stanleytown for production. The seventh characteristic is the relatively high variability in yields from the various manufacturing processes. The long manufacturing times, interplant dependencies, and the variable yields cause management to apply the buffer management technique to the assembly lines as the factory's constraints. As a result of management's use of the assembly lines as the constraints, management readily accepts the validity of the drum-buffer-rope method for planning and controlling production at Stanleytown.

Two additional characteristics also impact the MPS function by requiring the MPS to support an MRP calculation by extending the planning horizon beyond the stable 4.5-month time period. The eighth characteristic is the long lead time for raw lumber purchased from lumber mills. The ninth characteristic is the seasonality for the various species of raw logs used by Stanley. The long lead times and the seasonality of the logs cause the company to continue to use MRP to generate purchasing requirements by extending the MPS horizon for a full 12 months.

Priority Planning

The first key characteristic that impacts the priority planning function at Stanleytown's A-structure plants is the use of the assembly lines as the plants' constraints. The second characteristic is the relatively large number of end items that are made in the cabinet plant, chair plant, and table plant. The use of the assembly lines as constraints and the large number of end items mean that the plants do not need the due dates produced by the MRP system, as was required prior to the use of buffer management. As a result of the use of the Theory of Constraints rather than MRP for planning for the A-structure plants, the use of buffer management was readily accepted by management and production employees.

The third characteristic is the possibility of misallocating capacity in the A-structure plants. The fourth characteristic is the manufacturing time required for about 75% of the parts, which need less than seven work days. As a result of the potential for misallocation of material and the processing times for manufactured components, the drum-buffer-rope method establishes the actual, effective shop floor control system used in the plants, rather than an informal system. As a result of the acceptance of drum-buffer-rope as the shop floor control system, the key performance measurement used in the A-structure plants is the buffer due date status. As a result of the use of buffer management as the actual shop floor control system rather than an informal system, there is less work-in-process inventory than existed when MRP was used as the shop floor control system.

The fifth characteristic that impacts the priority planning function is the make-to-stock product environment, because the lumpy demand for final products triggers component requirements for specific wood species. The sixth characteristic is management's focus on delivery performance as the key measurement method for the manufacturing process. The make-to-stock environment and the criticality of delivery performance cause management to accept a relatively large amount of finished goods inventory in the finished goods warehouse. Priority planning to meet actual customer orders is accomplished through management of the finished goods stock. Additionally, the levels of finished goods trigger replenishment orders to the plants through the MPS function in a random fashion without respect to the capacity available.

Capacity Planning

The first critical characteristic that impacts the capacity planning function in the A-structure plants is the centralization of the MPS function. The second characteristic is the high product seasonality that must be leveled. The third characteristic is the five Stanley Company factories that are capable of producing specific product lines. As a result of the centralization of the MPS function, the end product seasonality, and the five production factories, the capacity planning function is an integrated part of the MPS process. Additionally, as a result of the integration with the MPS function, the capacity planning function is centralized.

The fourth characteristic that impacts capacity planning is the criticality of delivery performance. The fifth characteristic is management's view of capacity in terms of the amount of labor employed to support the assembly lines rather than the amount of equipment that is used in production areas. Most labor is multiskilled at Stanley. The use of delivery performance and the focus by management on the amount of labor causes capacity planning at Stanley to be seen as a function of the production rate of the assembly lines at the chair plant, table plant, and cabinet plant. As a result of management's use of the production rates, rather than standard hours or other capacity measures, the company does not use MRP output to analyze capacity in the A-structure plants. As a result of the use of the production rates to determine capacity in the A-structure plants, the manual calculations used for capacity planning give management acceptable results. An example of the manual calculations is if the assembly line rate is X, the current work force in fabrication is Y, and the line rate is changing to X + 1, then the work force should increase by the same ratio.

Priority Control

The first characteristic that impacts the priority control function is management's focus on the assembly lines as the constraints. The second characteristic is the use of the buffer report and FAS to schedule the convergent points in the production process in the A-structure plants. The focus on the assembly lines as constraints and the use of the buffer report cause the final assembly lines to be the drums for the A-structure plants. Because of the use of the assembly lines as the drums in each plant, the buffer schedule is the key priority control tool in the A-structure plants.

The third characteristic is the relatively large number of components required to be produced in the A-structure plants to support the assembly lines. The fourth characteristic is the limitation of the MRP system restricts the production requirements to weekly buckets and weekly lead times even though production on the assembly lines is scheduled to occur daily. The large number of components and the limitation of the MRP system combine with the use of buffer management to make MRP data insufficient for sched uling the shop. As a result of the lack of useful MRP scheduling data and the acceptable use of buffer management, the production areas in the A-structure plants readily accept the drum-buffer-rope scheduling method. However, because of the previous use of MRP, there is discomfort among some supervisors concerning the lack of intermediate due dates using the drum-buffer-rope method for nonconstraint feeding operations.

Capacity Control

The first characteristic that impacts the capacity control function in the A-structure plants is the buffers that are used to absorb the instabilities in the manufacturing processes. The second characteristic is the existence of idle capacity in nonconstraint operations. The ability of the buffers to dampen the unplanned instability of manufacturing and the idle capacity cause capacity control to be a manual process in the A-structure plants, where few adjustments are necessary.

The third characteristic is the use of the assembly lines as the constraints for the A-structure plants. The fourth characteristic is the use of the buffer status to determine capacity requirements in the short run. The use of the assembly lines as the constraints and the buffer position to determine the capacity requirements combine to cause management to use the FAS as the A-structure plants' drum. By using the assembly line as the constraint, limited use of overtime is required for manufacturing areas to support the buffer.

SUMMARY

The Stanley Furniture Company uses the drum-buffer-rope production planning and control method and the underlying Theory of Constraints to manage the production environment at the Stanleytown factory. The company supplies final customer demand from a finished goods warehouse and develops its MPS and FAS in order to replenish warehouse inventory. A significant total quality management program is also under way. The drum-buffer-rope method has generally replaced all shop floor control functions that had been performed by the MRP system. The MRP system is primarily used for planning and controlling purchase requirements. The focus of Stanley management is on the delivery of products to the customer in a timely manner and the reduction of manufacturing lead times to support the customer demand and reduce finished goods inventory. Theory of Constraints performance measurement methods are used in the factory to evaluate management actions.

OVERVIEW OF OTHER CONSTRAINTS MANAGEMENT TOOLS

III

PERFORMANCE MEASUREMENT

INTRODUCTION

> Tell me how you will measure me and I will tell you how I will perform. If you measure me in an illogical way...Do not complain about illogical behavior.
> *Goldratt (Theory of Constraints)*

In this chapter, organizational performance measurement is discussed. This topic deserves much more space than one chapter. Times have changed, but the way we manage and measure in organizations has not changed. Our body of knowledge has not changed to address the new competitive environment. Most managerial accounting, cost accounting, finance, budgeting, and engineering economics texts should be rewritten. The contents of these texts do not recognize:

- The constraint determines the throughput of the entire organization.
- The departments and processes of an organization form a chain from raw materials to customer. One must view this chain as a system, as a whole, and not manage and measure each link as if it were independent of the other links.
- There are relationships between one activity and another in an organization.
- Traditional measures of local performance no longer apply.
- The cost structure of most organizations has shifted significantly from variable to fixed.
- Product variety, high quality, short lead time, due date performance, etc. are competitive edges that must be linked to organizational profitability.

- Cost allocation distorts true costs.
- Storing costs in inventory leads to building excessive inventories without really increasing profits.
- Ultimately, the goal of a company is to make money, not to save money.

In this chapter, the problems created by the manufacturing performance measurement system are discussed. The current reality tree of the Goldratt Thinking Processes is introduced to demonstrate the causal connections from undesirable effects to core problems. Next, a simple Gedanken (thought) exercise is used to show the impact of traditional measures on decision making, and the Theory of Constraints measures of throughput, inventory, and operating expense are discussed. Finally, the current reality tree is used to communicate the impact of poor measures on the V-, A-, and T-plant structures. In this manner, you can more fully understand the causes of the day-to-day problems you face. These causes are the core problems that must be addressed to achieve continuous organizational improvement as a day-to-day activity. Finally, the steps in a process of ongoing improvement are again discussed.

WHY IS IMPROVEMENT SO DIFFICULT IN MOST ORGANIZATIONS?

After intense struggling, most companies make marginal progress in striving to reduce costs and in some cases increase revenues, only to face a new obstacle and stagnate or lose any gains. Why the struggle? Why only marginal progress? Why a new obstacle? Why stagnation or loss? Managers face significant resistance to change at all levels in the organization. Time and again, we see organization after organization surge forward with a competitive advantage only to stagnate and decline. Why?

Let's use the current reality tree (CRT) of Goldratt's Thinking Processes[1] to delve deeper into our understanding of the organization and its environment. The CRT is a "logic-based tool for using cause and effect relationships to determine root problems that cause the observed undesirable effects of the system" (p. 19).[2] Constructing a CRT is a simple but time-consuming, soul-searching process. You continually ask yourself *"why"*—why something exists and what causes it. In answering this question, you are expanding your understanding of the problem area. In the next chapter, the steps in constructing a CRT will be described. This chapter focuses on the use of the CRT in communicating the causal relationships in a situation. You will learn how to read a CRT.[3]

Let's start by reading the CRT in Figure 10.1 to broaden our understanding of why change is so difficult to implement in most organizations. The CRT is read from the bottom up using IF-THEN logic. Two types of logical connectors exist-simple and compound statements. IF (entity—tail of arrow)-THEN (entity—tip of arrow) logic is the simple connector, and IF (entity—tail of arrow)-AND (entity—tail of arrow)-THEN (entity—tip of arrow) logic is the compound connector. The second type of connector is called a *conceptual and* relationship. Both entities (tails of the arrows) must exist for the result (tips of the arrows) to exist. This sounds complex but, as in most new learning, it is simpler to learn through an example than through abstract instructions.

The entities at the bottom of the tree do not have arrows entering them. This type of entity is called a core driver. A core driver is a descriptor of the environment and influences the problem area but is beyond the control (or influence) of the tree builder or is an entity that the builder does not want to address for some reason. The second type of core driver is a core problem. According to Goldratt, a core problem is defined as the source of 70% or more of the undesirable effects in the tree. Let's read the entities at the bottom of the diagram to start the process. These entities have no arrows entering their blocks (the entities are numbered to help trace the logic): "5 Today, operations resources are quite expensive." "3 In organizations, local measures are needed to monitor daily operations performance." "1 Reducing product cost is viewed as the major goal of a traditional operations performance measurement system (PMS)." "10 Today, direct labor is quite expensive." Do you agree with these statements? There is near universal agreement with these statements among manufacturing professionals. Assuming that these statements are true in your reality, let's continue. The bar across the connecting arrows indicates the AND connector. IF "5 Today, operations resources are quite expensive" AND "3 In organizations, local measures are needed to monitor daily operations performance" AND "1 Reducing product cost is viewed as the major goal of a traditional operations PMS," THEN "12 Resource utilization is a local measure believed to be linked to costs (e.g., the higher the resource utilization, the lower the product cost"). Is this true in your environment? IF "5 Today, operations resources are quite expensive" AND "1 Reducing product cost is viewed as the major goal of a traditional operations PMS," THEN "15 Resource depreciation and associated expenses

102 Frequently, first-line supervisors ignore batch due dates.

58 Frequently first-line supervisors combine & sequence batches to save time on setups.

48 First-line supervisors are measured based on direct labor efficiency.

101 Process batch sizes are much larger than customer orders.

55 Sometimes operations managers release parts early to keep workers busy.

45 One way to reduce direct labor costs per part is to keep workers busy producing parts.

53 Sometimes operations managers release large batches of a part having no customer demand to keep resources busy.

44 One way to increase resource utilization is to produce large batches of parts.

100 There are many more batches on the shop floor than customer orders require.

50 Sometimes operations managers increase batch sizes to keep resources busy.

40 First-line supervisors are measured based on resource utilization.

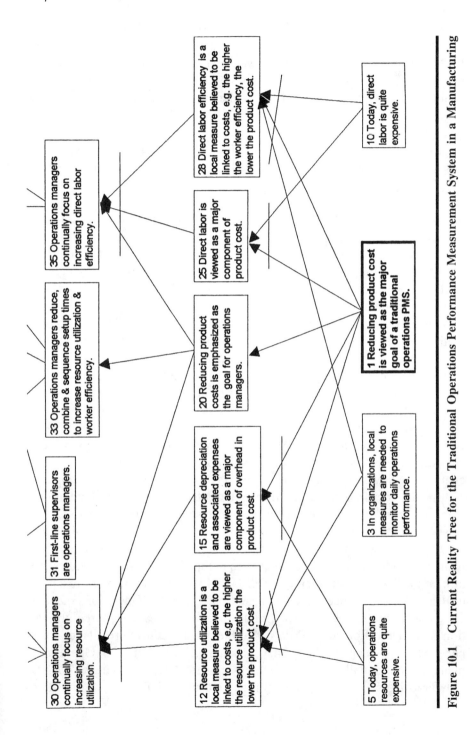

Figure 10.1 Current Reality Tree for the Traditional Operations Performance Measurement System in a Manufacturing

are viewed as a major component of overhead in product cost." Is this true? IF "1 Reducing product cost is viewed as the major goal of a traditional operations PMS," THEN "20 Reducing product costs is emphasized as the goal for operations managers." IF "1 Reducing product cost is viewed as the major goal of a traditional operations PMS" AND "10 Today, direct labor is quite expensive," THEN "25 Direct labor is viewed as a major component of product cost." IF "10 Today, direct labor is quite expensive" AND "3 In organizations, local measures are needed to monitor daily operations performance" AND "1 Reducing product cost is viewed as the major goal of a traditional operations PMS," THEN "28 Direct labor efficiency is a local measure believed to be linked to cost (e.g., the higher the worker efficiency, the lower the product cost)." Is this true in your organization?

IF "12 Resource utilization is a local measure believed to be linked to costs (e.g., the higher the resource utilization, the lower the product cost)" AND "15 Resource depreciation and associated expenses are viewed as a major component of overhead in product cost" AND "20 Reducing product costs is emphasized as the goal for operations managers," THEN "30 Operations managers continually focus on increasing resource utilization." True in your environment? IF "20 Reducing product costs is emphasized as the goal for operations managers," THEN "33 Operations managers reduce, combine & sequence setup times to increase resource utilization & worker efficiency." IF "20 Reducing product costs is emphasized as the goal for operations managers" AND "25 Direct labor is viewed as a major component of product cost" AND "28 Direct labor efficiency is a local measure believed to be linked to costs (e.g., the higher the worker efficiency, the lower the product cost)," THEN "35 Operations managers continually focus on increasing direct labor efficiency." IF "30 Operations managers continually focus on increasing resource utilization" AND "31 First-line supervisors are operations managers," THEN "40 First-line supervisors are measured based on resource utilization." IF "33 Operations managers reduce, combine & sequence setup times to increase resource utilization & worker efficiency," THEN "44 One way to increase resource utilization is to produce large batches of parts." IF "33 Operations managers reduce, combine & sequence setup times to increase resource utilization & worker efficiency," THEN "45 One way to reduce direct labor costs per part is to keep workers busy producing parts." IF "35 Operations managers continually focus on increasing direct labor efficiency" AND "31 First-line supervisors are operations managers," THEN "48 First-line supervisors are measured based on direct labor efficiency." IF "30 Operations managers continually focus on increasing resource utilization" AND "44 One way to increase resource utilization is to produce large batches of parts," THEN "50 Sometimes operations managers increase batch sizes to keep resources busy." IF "30 Operations managers continually focus on increasing resource utilization" AND "44 One way to increase resource utilization is to produce large batches of parts," THEN "53 Sometimes operations managers release large batches of a part having no customer demand to keep resources busy." IF "35 Operations managers continually focus on increasing direct labor efficiency" AND "45 One way to reduce direct labor costs per part is to keep workers busy producing parts," THEN "55 Sometimes operations managers release parts early to keep workers busy." IF "33 Operations managers reduce, combine & sequence setup times to increase resource utilization & worker efficiency" AND "48 First-line supervisors are measured based on direct labor efficiency," THEN "58 Frequently first-line supervisors combine & sequence batches to save time on setups." Does this argument sound logical to you? Does it exist in your facility? Let's complete the argument.

IF "50 Sometimes operations managers increase batch sizes to keep resources busy," THEN "100 There are many more batches on the shop floor than customer orders require." IF "53 Sometimes operations managers release large batches of a part having no customer demand to keep resources busy," THEN "100 There are many more batches on the shop floor than customer orders require." IF "55 Sometimes operations managers release parts early to keep workers busy," THEN "100 There are many more batches on the shop floor than customer orders require." IF "50 Sometimes operations managers increase batch sizes to keep resources busy," THEN "101 Process batch sizes are much larger than customer orders." IF "58 Frequently first-line supervisors combine & sequence batches to save time on setups," THEN "101 Process batch sizes are much larger than customer orders." IF "58 Frequently first-line supervisors combine & sequence batches to save time on setups," THEN "102 Frequently, first-line supervisors ignore batch due dates." The argument is complete. Does it sound logical? Is this what you see—high inventories and late orders?

Does your facility have many more batches, large batches (many times greater than customer orders), and/or batches that have no customer demand on the shop floor? Does it seem as though first-line supervisors ignore batch due dates?

Let's study the diagram carefully. We find that two characteristics of today's competitive environment drive many of the actions in the CRT. These characteristics of the environment are:

- Today, operations resources are quite expensive. (1/5)*
- Today, direct labor is quite expensive. (1/10)

Additionally, there are characteristics that define our traditional operations performance measurement system. These characteristics include:

- In organizations, local measures are needed to monitor daily operations performance. (1/3)
- Reducing product cost is viewed as the major goal of a traditional operations PMS. (1/1)
- Local measures are assumed to link causally to global measures. (basic assumption)

These characteristics of the environment and the PMS cause operations managers to take specific actions to improve their measures. These actions include:

- Sometimes operations managers increase batch sizes to keep resources busy. (1/50)
- Sometimes operations managers release large batches of each part having no customer demand to keep resources busy. (1/53)
- Sometimes operations managers release parts early to keep workers busy. (1/55)
- Frequently first-line supervisors combine & sequence batches to save time on setups .(1/58)
- Operations managers continually focus on increasing resource utilization. (1/30)
- Operations managers continually focus on increasing direct labor efficiency. (1/35)

The end results from these manager actions are:

- There are many more batches on the shop floor than customer orders require. (1/100)
- Process batch sizes are much larger than customer orders. (1/101)
- Frequently, first-line supervisors ignore batch due dates. (1/102)

---

* Numbers in parentheses indicate figure number and entity number.

These results have significant impact on meeting the customers' needs with respect to products, variety, lead times, due dates, quality, etc. Let's stop here for now. We will link these results to V-, A-, and T-plant structures later in the chapter.

The core problem of this CRT analysis is: "**1 Reducing product cost is viewed as the major goal of a traditional operations PMS.**" This is the driver that created excessive inventories on the shop floor and caused the first-line supervisor to ignore batch due dates. To eliminate the undesirable effects, we must address this core problem. We must not only find an appropriate measurement system to replace the focus on reducing product cost, but we must also change the organizational culture from one where everyone tries to look busy when the boss walks through the area to one where worker idleness is viewed as an opportunity to look for organizational process improvements.

Remember that the operations area is only one of three major segments of any organization. What of the other two segments? The second segment, the customer functions, consists of marketing, sales, and field service. These functions have close contact with the organization's customers in marketing and selling to customers and servicing the customers' needs. If we studied this customer segment using a CRT, we would find their core problem to be: "**1 Increasing revenues is viewed as the major goal of a traditional customer function PMS.**" The third segment, the finance segment, consists of the accounting and finance functions. This segment reviews the investment proposals of the other two segments and measures their impact on short-term costs and revenues using traditional cost and managerial accounting methods. The assumption of these traditional methods is that if we review costs and select a local cost-savings proposal, then the organizational profit is increased. In contrast, if we review revenue-generating proposals and select a local revenue-generating proposal, then the results translate into more organizational profit. Traditional cost accounting and managerial accounting do not recognize the importance of the constraint in determining organizational profit; nor do they recognize the dependencies across resources and functions; nor do they recognize that organizational cost structures have changed significantly since the development of cost and managerial accounting in the early 1900s. Additionally, in comparing various proposals, the finance segment focuses on short-term financial gains. If we studied this segment, we would find two additional core problems. The first problem is: "**1 Local improvement in the resource segment and in the customer** segment is believed to translate directly into organizational improvement." The second is: "**2 A short-term financial focus is required in an organization.**"

The end product of this traditional business strategy is the belief that IF the operations functions focus on local cost-savings opportunities AND the customer functions focus on local revenue-generation opportunities AND the finance functions focus on short-term gains in costs and revenues, THEN the organization will make *long-term profits*. This belief is false, but our traditional management philosophy is built on this foundation. Lockamy and Cox[8] provide a detailed discussion of this problem and its solution in *Reengineering Performance and Measurements*. Let's study a simple example to illustrate the problems of traditional management.

BOB'S BOLT COMPANY—REVISITED

Bob has a small manufacturing plant that makes four simple products—A, B, C, and D. These products sell for $110, $110, $120, and $125, respectively. Weekly demand is 30 units for A, 30 units for B, 40 units for C, and 40 units for D. The plant has three production work centers (WC X, WC Y, and WC Z) and three production workers, one assigned to each work center (workers are specialists and cannot be transferred from one work center to another). Workers are paid $6 per hour; overhead is 200% of direct labor costs. Each worker works 40 hours per week (2400 minutes). Raw material a is used to make product A and costs $50 per unit, raw material b is used to make product B and costs $55 per unit, raw material c is used to make product C and costs $60 per unit, and raw material d is used to make product D and costs $60 per unit. One unit of the appropriate raw material is used in each product.

Next, let's look at the product routings. The routing for product A is raw material a moves to WC Z, then WC Y, then WC X. In WC Z the operation time per part is 30 minutes, in WC Y the operation time is 30 minutes per unit, and in WC X the operation time is 30 minutes per unit. The routing for product B is raw material b moves to WC Z, then WC Y, then WC X. In WC Z the operation time per part is 30 minutes, in WC Y the operation time is 30 minutes per unit, and in WC X the operation time is 30 minutes per unit. The routing for product C is raw material c moves to WC Z, then WC Y, then WC X. In WC Z the operation time per part is 10 minutes, in WC Y the operation time is 45 minutes per unit, and in WC X the operation time is 10 minutes per unit. The routing for product D is raw material d moves to WC Z, then WC Y, then WC X. In WC Z the operation time per part is 20 minutes, in WC Y the operation time is 40 minutes per unit, and in WC X the operation time is 10 minutes per unit. The Gedanken exercise information is summarized in Figure 10.2 and Table 10.1.

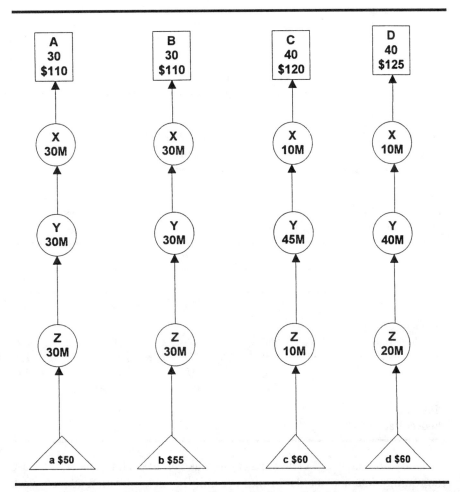

Figure 10.2　Bob's Bolt Company Product Information (three workers [one at each WC X, Y, and Z], each worker earns $6 per hour and works 40 hours per week; the overhead rate is 200% of direct labor)

Table 10.1 Bob's Bolt Company Information

	Products			
	A	*B*	*C*	*D*
Demand	30	30	40	40
Price	$110	$110	$120	$125
Routing/Standards				
X	30M	30M	10M	10M
Y	30M	30M	45M	40M
Z	30M	30M	10M	20M
Raw Materials	a $50	b $55	c $60	d $60

Let's use traditional cost accounting to determine which products to make, their quantities, and Bob's estimated profit (Table 10.2). Based on using product cost and product profit as the measure for determining the product mix of Bob's Bolt Company, Bob would make product D first, C second, A third, and B fourth.

Table 10.2 Bob's Bolt Company—Solution Based on Traditional Cost Accounting Using Product Profit

	Products							
		A		*B*		*C*		*D*
Product Price		$110		$110		$120		$125
Product Cost								
Raw Materials	a	$50	b	$55	c	$60	d	$60
Direct Labor (DL)	90M	9	90M	9	65M	6.50	70M	7.00
Overhead (200% DL)		18		18		13		14
		$77		$82		$79.50		$81
Product Profit		$33		$28		$40.50		$44
Product Rank (Make)		3		4		2		1

Does Bob have sufficient capacity to make all of each product? Let's use the routing data and standard times per unit for each operation at each work center to determine the load placed on each work center (Table 10.3). We are conducting the first step of the Theory of Constraints five-step focusing process for continuous improvement. **Step 1: Identify the system constraint.**

Each work center's required capacity to meet market demand is greater than 100%. In this case, the constraint is WC Y, the heaviest loaded work center, with a 216.67% load when comparing capacity required to capacity available. We will ration our capacity of 2400 minutes at WC Y to those products having the highest product profit in descending order. The results are shown in Table 10.3.

Table 10.3 Bob's Bolt Company—Capacity Analysis Based on Market Demand

	Product			
	A	B	C	D
Demand	30	30	40	40
WC X	30M 30(30M) + 2600M/2400M × 100% = 108.33%	30M 30(30M) +	10M 40(10M) +	10M 40(10M) =
WC Y*	30M 30(30M) + 5200M/2400M × 100% = 216.67%	30M 30(30M) +	45M 40(45M) +	40M 40(40M) =
WC Z	30M 30(30M) + 3000M/2400M × 100% = 125%	30M 30(30M) +	10M 40(10M) +	20M 40(20M) =

The market demand of 40 units of D per week consumes 1600 minutes of capacity, leaving 800 minutes (2400 minutes minus 1600 minutes). These 800 minutes can be assigned to making product C at 45 minutes per unit. Approximately 17.8 units can be made each week before exhausting the 2400 minutes available per week. Bob should make 40 Ds and 17.8 Cs each week based on using product profit to determine his product mix. Calculations are provided in Table 10.4.

Table 10.4 Bob's Bolt Company—Determining How Many to Make of Each Product

	Products			
	A	B	C	D
Demand	30	30	40	40
Product Rank (Make)	3	4	2	1
WC Y*	30M	30M	45M	40M
Capacity Required			800M +	1600M
Make	0	0	17.8	40

With this mix, how much profit can Bob make for his company? The calculations are provided in Table 10.5. Bob can make $1508 per week by making this product mix.

Table 10.5   Bob's Bolt Company—Organization Profit Based on Product Profit

		Product		
	A	*B*	*C*	*D*
Demand	30	30	40	40
Product Price	$110	$110	$120	$125
Raw Materials	a $50	b $55	c $60	d $60
Contribution	$60	$55	$60	$65
Make	0	0	17.8	40
Total Contribution	0	0	1068	+ 2600 =
				$3668
Direct Labor (3 workers × $6/hr × 40 hr/wk) = $ 720				
Overhead		1440		
Operating Expense				$2160
Organization Profit				$1508

Let's check the capacity associated with producing this product mix to insure we are not violating Bob's capacity available at each work center. These calculations are shown in Table 10.6.

Table 10.6   Bob's Bolt Company—Capacity Check for Using Product Mix Based on Product Profit

		Product		
	A	*B*	*C*	*D*
Make	0	0	17.8	40
WC X	30M	30M	10M	10M
	0(30M) + 578M/2400M × 100%= 24%	0(30M) +	17.8(10M) +	40(10M) =
WC Y*	30M	30M	45M	40M
	0(30M) + 2400M/2400M × 100%=100 %	0(30M) +	17.8(45M) +	40(40M) =
WC Z	30M	30M	10M	20M
	0(30M) + 978M/2400M × 100%= 41%	0(30M) +	17.8(10M) +	40(20M) =

As can be seen from the above capacity analysis, WC Y is fully loaded at 100% while both WCs X and Z are well below 100% utilization.

Another approach to determining product mix that is frequently used by companies is the contribution per direct labor hour (in our case minute). Let's see what the product mix is using this approach to prioritize products.[4,5] The calculations are shown in Table 10.7.

Table 10.7   Bob's Bolt Company—Solution Based on Traditional Cost Accounting Using Contribution per Direct Labor Unit

	Product			
	A	*B*	*C*	*D*
Demand	30	30	40	40
Product Price	$110	$110	$120	$125
Raw Materials	a $50	b $55	c $60	d $60
Contribution	$60	$55	$60	$65
Direct Labor (DL)	90M	90M	65M	70M
Contribution/DL Min.	$0.67	$0.61	$0.92	$0.93
Rank based on Cont./DL Min.	3	4	2	1

These rankings are identical to those given in Table 10.2 (ranking based on product profit).

In the traditional approach to determining product mix, we had to identify the constraint to calculate how many of each product could be made. Before starting with our constraint management calculations, let's review the performance measures and the five focusing steps of continuous improvement.[4] The definitions are provided again in Figure 10.3 (given previously in Figure 1.5). Review them carefully before proceeding.

---

five focusing steps—In the theory of constraints, a process to continuously improve organizational profit by evaluating the production system and market mix to determine how to make the most profit using the system constraint. The steps consist of 1) identifying the constraint to the system, 2) deciding how to exploit the constraint to the system, 3) subordinating all non constraints to the constraint, 4) elevating the constraint to the system, 5) returning to step 1 if the constraint is broken in any previous step, while not allowing inertia to set in (p. 31).[2]

throughput—In theory of constraints, the rate at which the system (firm) generates money through sales (p. 85).[2]

Figure 10.3   Constraints Management Performance Measurement Terms. From Cox, J.F., J.H. Blackstone, and M.S. Spencer, *APICS Dictionary*. Fall Church, VA: American Production and Inventory Control System, 1995. With permission.

inventory—In theory of constraints, inventory is defined as those items purchased for resale and includes finished goods, work in process and raw materials. Inventory is always valued at purchase price and includes no value-added costs, as opposed to the traditional cost accounting practice of adding direct labor and allocating overhead as work in process progresses through the production process (p. 40).[2]

operating expense—In theory of constraints, the quantity of money spent by the firm to convert inventory into sales in a specific time period (p. 55).[2]

Figure 10.3   Constraints Management Performance Measurement Terms (continued)

We have previously completed **Step 1: Identify the system constraint.** WC Y was identified as a system constraint. **Step 2: Decide how to exploit the system constraint** means that once the constraint is identified, we should make the most from its use. Where do we make the most money from the use of WC Y? Let's use a product profit calculation similar to that in Table 10.7 but focus on the effective use of WC Y, our system constraint. The calculations are shown in Table 10.8.

Table 10.8   Bob's Bolt Company—Solution Based on Theory of Constraints Measures and Focusing Steps

		Product		
	A	B	C	D
Demand	30	30	40	40
Product Price	$110	$110	$120	$125
Raw Materials	a $50	b $55	c $60	d $60
Contribution	$60	$55	$60	$65
WC Y*	30M	30M	45M	40M
Contribution/DL Min.	$2	$1.83	$1.33	$1.62
Rank (based on WC Y use)	1	2	4	3

Based on effectively utilizing the constraint, WC Y, the ranking for Bob's product mix is quite different when compared to either traditional cost accounting method. Let's determine how many of each product can be made based on this constraint mix. The calculations are provided in Table 10.9.

The product mix of Bob's Bolt Company based on exploiting the constraint is 30 As, 30 Bs, and 15 Ds (**Step 2: Decide how to exploit the system constraint**). Let's now determine if this mix provides a higher product profit

Table 10.9 Bob's Bolt Company—Determining How Many to Make of Each Product

	Products			
	A	B	C	D
Demand	30	30	40	40
Product Rank (Make)	1	2	4	3
WC Y*	30M	30M	45M	40M
Capacity Required	900M	900M	0M	+ 600M
Make	30	30	0	15

than the previous mix based on traditional accounting methods. The calculations are provided in Table 10.10.

Table 10.10 Bob's Bolt Company—Organization Profit Based on Constraint Management Contribution per Constraint Unit

	Product			
	A	B	C	D
Demand	30	30	40	40
Product Price	$110	$110	$120	$125
Raw Materials	a $50	b $55	c $60	d $60
Contribution	$60	$55	$60	$65
Make	30	30	0	15
Total Contribution	1800	1650	0	+ 975 = $4425

Direct Labor (3 workers × $6/hr × 40 hr/wk) = $ 720
Overhead 1440
Operating Expense $2160
Organization Profit $2265

Before discussing the ramifications of this approach, let's check the capacity of each work center for this mix. This calculation is provided in Table 10.11. Capacity is ample at all work centers for the constraint product mix. Now, let's examine what is causing the significant increase in profit. Profit increased from $1508 to $2265, over 50%, by focusing on the mix that effectively used the constraint capacity.

Table 10.11 Bob's Bolt Company—Capacity Check for Using Product Mix Based on Product Profit

	Product			
	A	B	C	D
Make	30	30	0	15
WC X	30M	30M	10M	10M
	30(30M) +	30(30M) +	0(10M) +	15(10M) =
	1950M/2400M × 100%= 81%			
WC Y*	30M	30M	45M	40M
	30(30M) +	30(30M) +	0(45M) +	15(40M) =
	2400M/2400M × 100%=100 %			
WC Z	30M	30M	10M	20M
	30(30M) +	30(30M) +	0(10M) +	15(20M) =
	2100M/2400M × 100% = 87.5%			

Let's continue with Bob's Bolt Company to illustrate Steps 3 thorough 5 of the five focusing steps. **Step 3: Subordinate all else to the constraint of the system.** What if WC Z is allowed to work at 100% utilization. The first-line supervisor would probably get more materials released to the shop floor than required to support the desired product mix. Any raw material release would probably help improve the utilization of WC Z. But what happens when more raw materials are released for any product A, B, C, or D? There are no current orders for the excess. WC Z works on the raw materials and sends all work-in-process to WC Y. WC Y is the constraint and determines the throughput and profit for the organization. What if WC Y works on the wrong product? Bob loses the sales for that week—profit lost is profit lost forever. A nonconstraint work center such as WC Z working to full utilization only creates excess WIP on the shop floor and increases the probability of other work centers working on increasing inventory instead of throughput. **Subordinate means just that—work at the pace to support the constraint capability and mix.**

Step 4: Elevate the constraint of the system. Buy more of the constraint capability. This is the approach to increasing throughput once everything else has been synchronized to the constraint capability. This elevation should be a conscious thoughtful decision. Overtime, second or third shifts, working through lunch and breaks, subcontracting, setup improvement, maintenance, etc. are approaches to increasing the throughput of the constraint. Buying a new piece of equipment is the obvious option. Most efforts to improve nonconstraints will not increase throughput significantly, if any. Efforts to improve nonconstraints may reduce costs. Make sure you know what the real impact of capital improvement efforts are. Most traditional evaluation approaches provide incorrect evaluation.

Step 5: If in step 4 the constraint is broken, go to step 1, but do not allow inertia to set in. Occasionally in step 4 the constraint is broken—the constraint moves from its current location to another location or even to another function (marketing, sales, engineering, purchasing, etc.). In most instances when the constraint moves from the operations function to another function, the new constraint is a management policy. It may be a policy such as sales commission based on product price or margin or not selling below product cost plus X% margin or not selling to the government, globally, etc. It may be that the organization does not sell in a generic package, under private label, to competitors, in the spare parts market, etc. There are literally thousands of policies, both written and unwritten, that could block future throughput. Here, each company is somewhat unique. The CRT of the Thinking Processes is an excellent tool for identifying the true core problems in a situation. It is the only tool that we know of for surfacing policy constraints. It is the major topic of the next chapter. For now, let's wrap up the measurement issues related to Bob's Bolt Company.

Do you use cost accounting information to determine product prices, to determine which products to focus sales efforts on, to determine which products to discontinue, to determine your make versus buy decisions, to determine what to charge sister plants for a product (transfer pricing), to determine sales commissions, and a myriad of other organizational decisions?

Think about what you experienced in examining Bob's Bolt Company. You probably had a sinking feeling in the pit of your stomach. How many of those decisions were poor ones. How many products did you discontinue that were probably excellent products based on your resource capabilities? Look at Tables 10.2 and 10.7 carefully. With both traditional methods, you probably would have recommended discontinuing products A and B as they certainly did not rank high in contributing to organizational profit based on traditional accounting methods. Cull out the "dogs." And yet they make the highest organizational profit. Your company probably doesn't have four products and three work centers. Your company may have dozens of each. What is the probability that your company is making the right products and making the right decisions based on your resources?

How many times have you purchased new equipment based on cost accounting data? How many times did the benefits really materialize? There is only one place where significant throughput results in our example and in most companies. Do you look at constraints to determine improvement opportunities? How many parts do you buy from outside vendors when you have nonconstraint capacity? Your make versus buy decisions certainly warrant further analysis if you used traditional product cost data. How do you price your products? Do you use a cost roll up and add a profit margin? How much of that cost is direct labor and allocated overhead? How many products could you sell for below product cost and still make significant organizational profit? More business has been turned away because of product costing than you might imagine? Think about how much of the cost is arbitrary and the organization will incur anyway, whether it makes the product or not.

A few hours spent studying this example and the types of decisions your organization makes will open your eyes to just how bad the decisions you have been making based on your traditional accounting system really are. At the beginning of this chapter, we said that one chapter was not enough to cover the issues related to performance measurement. Our motivation for making that statement should now be clear. Performance measurement provides the foundation for good business. Today, cost accounting principles and the focus on reducing product cost provide the foundation for traditional business philosophy and practice. In the authors' opinion, this area is wrong on both counts.

WHY IS IMPROVEMENT SO DIFFICULT IN MOST ORGANIZATIONS?—REVISITED

Let's revisit our earlier discussion of why it is so difficult to improve in most organizations. We saw the impact of using the traditional performance measurement system on an organization in Figure 10.1 and in the Gedanken exercise. We found that most capital improvement efforts lead to nothing more than increased idle time at nonconstraints. Seldom was the true constraint identified and exploited to gain the most money. We found that policies and procedures were the biggest obstacles to true bottom-line improvement. We found that each link in the chain is forced to act independently of the other links, and measures across functions cause conflicts among the functions and less than desirable organizational results. We found that most measures lead us to flood the shop floor with larger and larger batches and cause first-line supervisors to ignore due dates for parts. The end results of these efforts are higher and higher efficiencies and utilizations, until the customers start calling about their late orders. Customer complaints and complaints from salesmen and field service representatives create chaos. Salesmen and field service personnel see their commissions dwindling. In the plant, setups were broken, batches were split, parts were expedited through the system, overtime was used at key work centers, and premium freight was incurred to get the products to the customers quickly. All of these actions reduced the efficiencies and utilizations of the operations managers—their performance measures. They were heroes for a moment for getting the order out, but their reaction after the chaos settled down was to regain the lost production time through larger batches, sequencing, combining batches, and ignoring batch due dates. Any manager who was unable to increase efficiencies and utilizations after the chaos found his or head head on the chopping block at the end-of-the-month performance review meeting. The CRT showed the core problem, the policies and procedures that management implemented based on using a traditional performance management system, and the organization's environment. From these entities, the actions of managers were identified, as were the results of these actions. These actions and results have become a way of life—the culture—for most manufacturers. Many people might think that this situation is enough of an obstacle in an organization to prevent true progress. Let's not stop here. This is only part of the problem. Let's fully expose the core problems to true organizational improvement.

The errors of traditional cost and managerial accounting are illustrated next through a simple Gedanken exercise. We found that the decisions concerning which products to produce and sell are usually based on cost accounting calculations of product cost and product profit. We saw that this approach can lead to large losses in organization profit. Planning based on the local optima does not provide a plan to produce the global optimum.

Thus far, we have identified the development of a poor production plan—a poor product mix that does not exploit the capabilities of the organization. We also have identified measures that block the execution of any plan. These are two of the reasons why it is so difficult to implement a process of continuous improvement. There are more reasons, however. Let's look at the underlying assumption of a traditional planning and control system based on our understanding of constraints management.

Characteristics of a Traditional Scheduling and Control System

Today, most companies have or are implementing material requirements planning (MRP) logic based information systems. These systems utilize a bill of material file and a routing file to determine the materials and capacity needs to complete the master production schedule. The *APICS Dictionary*[2] (pp. 49-50) defines MRP as "A set of techniques that uses bill of material data, inventory data, and the master production schedule to calculate requirements for materials. It makes recommendations to release replenishment orders for material. Further, because it is time-phased, it makes recommendations to reschedule open orders when due dates and need dates are not in phase. Time-phased MRP begins with the items listed on the MPS and determines (1) the quantity of all components and materials required to fabricate those items and (2) the date that the components and material are required. Time-phased MRP is accomplished by exploding the bill of material, adjusting for inventory quantities on hand or on order, and offsetting the net requirements by the appropriate lead times." Manufacturing resource planning (MRPII) and enterprise resource planning (ERP) are more sophisticated versions of MRP using the same logic and policies and procedures as MRP. As computer capabilities have increased a thousand-fold over the last 30 years, no one except Goldratt has challenged the underlying logic of our traditional policies and procedures embedded in these systems. Goldratt, with his OPT software (popular in the early 1980s), was the first to challenge the underlying logic of traditional scheduling and control systems. In 1990, in *The Haystack Syndrome: Sifting Information Out of the Data Ocean,*[6] Goldratt provided the specifications for a scheduling and control system built upon planning, scheduling, and controlling the resource constraint to a production system. Several forward finite schedulers have appeared on the market touting the benefits of constraints management; however, few software vendors really understand the full significance of the term "constraints management" and the significant differences Goldratt envisioned in operating an organization. Being forward finite schedulers, many still have the characteristics of the traditional scheduling and control system (listed below) built into them.

The characteristics of a traditional scheduling and control system are:

- Rules do not recognize resource dependencies. (4/213)*
- Work centers (WCs) are scheduled by order due date. (5/211)
- WCs are block scheduled. (4/212)
- WCs are loaded to infinite capacity. (4/211)

---

* Numbers in parentheses indicate figure number and entity number.

- All WCs are scheduled in detail. (basic assumption of MRP — Daily Dispatch List)
- A batch is processed at one WC before being moved to the next WC. (4/210)
- Fabrication and assembly are treated as separate plants. (6/268)
- Fabrication is based on making batches to forecast. (6/211)
- Assembly is scheduled initially based on customer due date. (6/210)

Read this list carefully to determine which scheduling policies and procedures apply to your facility. Some apply to scheduling in an A-plant, some in a V-plant, and some in a T-plant. There are many more scheduling policies and procedures imbedded in the software. Many probably create severe problems in your environment without your knowing it. Only a few policies and procedures have been listed here to demonstrate the impact they have on managing your facility. A CRT of each plant type is provided to show the causal linkages from the performance measurement system, from the traditional scheduling and control system, and from the specific plant characteristics to the undesirable effects that you probably experience in your organization.

A-Plant Current Reality Tree

The characteristics of an A-plant are:

- Parts routings are dissimilar. (4/220)
- Parts are unique to specific products. (4/223)
- Many parts are combined at convergent points (assembly areas) to make each product. (4/222)
- General-purpose equipment is used to perform different operations on different parts. (4/221)
- General-purpose equipment requires long setups for dissimilar operations. (4/250)

Let's see how these characteristics combine with the characteristics of the traditional production scheduling and control system (SCS) and the effects of the performance measurement system CRT to create the A-plant environment.* Start at the bottom of the CRT in Figure 10.4. "210 In a SCS, a batch

---

* The CRT does not include the causal relationships of high WIP and finished goods inventories on poor quality. This casuality has been long recognized by total quality management and just-in-time advocates.

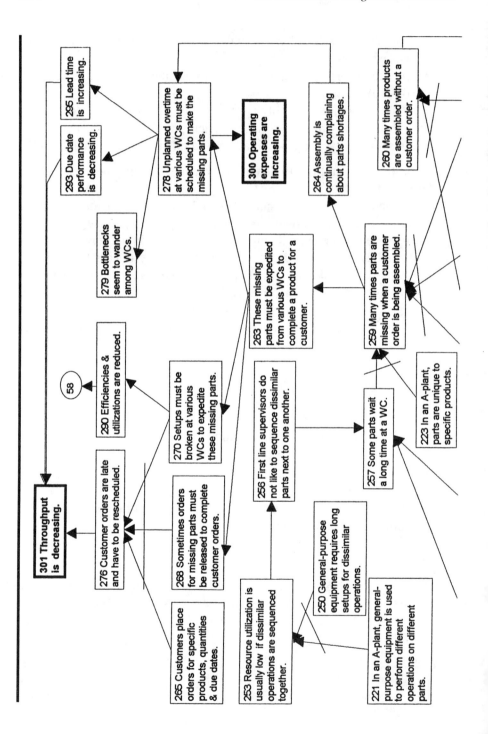

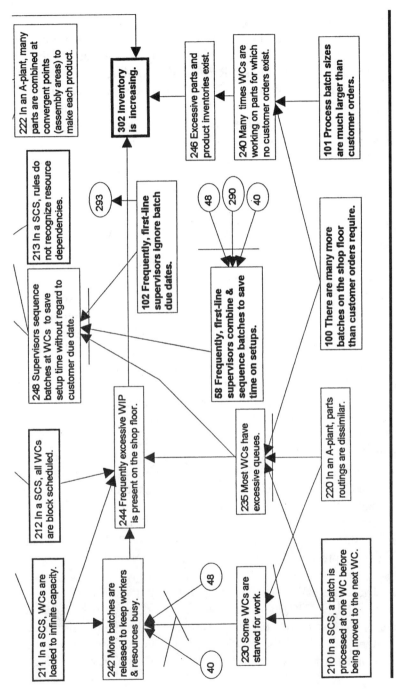

Figure 10.4   Current Reality Tree for a Traditional Scheduling System and Measures in an A-Plant

is processed at one WC before being moved to the next WC." "220 In an A-plant, parts routings are dissimilar." "100 There are many more batches on the shop floor than customer orders require." "101 Process batch sizes are much larger than customer orders." Recall "40 First-line supervisors are measured based on resource utilization," "48 First-line supervisors are measured based on direct labor efficiency," "58 Frequently first-line supervisors combine & sequence batches to save time on setups," and "102 Frequently, first-line supervisors ignore batch due dates" from the performance measurement system CRT (Figure 10.1). IF "210 In a SCS, a batch is processed at one WC before being moved to the next WC" AND "220 In an A-plant, parts routings are dissimilar," THEN "230 Some WCs are starved for work." IF "230 Some WCs are starved for work" AND "40 First-line supervisors are measured based on resource utilization," THEN "242 More batches are released to keep workers & resources busy." IF "230 Some WCs are starved for work" AND "48 First-line supervisors are measured based on direct labor efficiency," THEN "242 More batches are released to keep workers & resources busy." IF "242 More batches are released to keep workers & resources busy," THEN "244 Frequently excessive WIP is present on the shop floor." IF "210 In a SCS, a batch is processed at one WC before being moved to the next WC" AND "220 In an A-plant, parts routings are dissimilar" AND "100 There are many more batches on the shop floor than customer orders require," THEN "235 Most WCs have excessive queues." IF "235 Most WCs have excessive queues," THEN "244 Frequently excessive WIP is present on the shop floor." IF "211 In a SCS, WCs are loaded to infinite capacity," THEN "242 More batches are released to keep workers & resources busy." IF "211 In a SCS, WCs are loaded to infinite capacity," THEN "244 Frequently excessive WIP is present on the shop floor." IF "212 In a SCS, all WCs are block scheduled (scheduling using static rules, e.g., three days queue time at WC X, one day move time between work centers, two days inspection time, etc.)," THEN "244 Frequently excessive WIP is present on the shop floor." These logic paths lead to: IF "244 Frequently excessive WIP is present on the shop floor," THEN **"302 Inventory is increasing."**

IF "100 There are many more batches on the shop floor than customer orders require" THEN "240 Many times WC are working on parts for which no customer orders exist." IF "101 Process batch sizes are much larger than customer orders," THEN "240 Many times WCs are working on parts for which no customer orders exist." IF "240 Many times WCs are working on parts for which no customer orders exist," THEN "246 Excessive parts and product inventories exist." IF "246 Excessive parts and product inventories exist," THEN **"302 Inventory is increasing."**

IF "58 Frequently, first-line supervisors combine & sequence batches to save time on setups" AND "235 Most WCs have excessive queues" AND "102 Frequently, first-line supervisors ignore batch due dates," THEN "248 Supervisors sequence batches at WCs to save setup time without regard to customer due date." IF "248 Supervisors sequence batches at WCs to save setup time without regard to customer due date" AND "213 In a SCS, rules do not recognize resource dependencies" AND "222 In an A-plant, many parts are combined at convergent points (assembly areas) to make each product," THEN "259 Many times part are missing when a customer order is being assembled." IF "248 Supervisors sequence batches at WCs to save setup time without regard to customer due date" AND "222 In an A-plant, many parts are combined at convergent points (assembly areas) to make each product," THEN "260 Many times products are assembled without a customer order." IF "260 Many times products are assembled without a customer order," THEN **"302 Inventory is increasing."**

IF "211 In a SCS, WCs are loaded to infinite capacity," THEN "257 Some parts wait a long time at a WC." IF "212 In a SCS, all WCs are block scheduled," THEN "257 Some parts wait a long time at a WC." IF "221 In an A-plant, general-purpose equipment is used to perform different operations on different parts" AND "250 General-purpose equipment requires long setups for dissimilar operations," THEN "253 Resource utilization is usually low if dissimilar operations are sequenced together." IF "253 Resource utilization is usually low if dissimilar operations are sequenced together," THEN "256 First-line supervisors do not like to sequence dissimilar parts next to one another." IF "256 First-line supervisors do not like to sequence dissimilar parts next to one another," THEN "257 Some parts wait a long time at a WC." IF "257 Some parts wait a long time at a WC" AND "233 In an A-plant, parts are unique to specific products," THEN "259 Many times part are missing when a customer order is being assembled." IF "259 Many times part are missing when a customer order is being assembled," THEN "264 Assembly is continually complaining about parts shortages." IF "264 Assembly is continually complaining about parts shortages," THEN "278 Unplanned overtime at various WCs must be scheduled to make the missing parts." IF "259 Many times part are missing when a customer order is being assembled," THEN "263 These missing parts must be expedited from various WCs to complete a product for a customer." IF "263 These missing parts must be expedited from various WCs to complete a product for a customer," THEN "268 Sometimes orders for missing parts must be released to complete customer orders." IF "263 These missing parts must be expedited form various WCs to complete a product for a customer," THEN "270 Setups must be broken at various WCs to expedite these missing parts." IF "263 These missing parts must be expedited from various WCs to complete a product for a customer," THEN "278 Unplanned overtime at various WCs must be scheduled to make the missing parts." IF "278 Unplanned overtime at various WCs must be scheduled to make the missing parts," THEN **"300 Operating expenses are increasing."**

IF "270 Setups must be broken at various WCs to expedite these missing parts," THEN "290 Efficiencies & utilizations are reduced." IF "290 Efficiencies & utilizations are reduced" AND "40 First-line supervisors are measured based on resource utilization" (look below to 58) AND "48 First-line supervisors are measured based on direct labor efficiency," THEN "58 Frequently, first-line supervisors combine & sequence batches to save time on setups." Notice we now have a vicious loop in our tree. The loop is 58-248-259-263-270-290-58. A loop in a CRT mean that the situation keeps getting worse and worse as time passes. Is this the case in your facility?

IF "265 Customers place orders for specific products, quantities & due dates" AND "268 Sometimes orders for missing parts must be released to complete customer orders" AND "270 Setups must be broken at various WCs to expedite these missing parts," THEN "276 Customer orders are late and have to be rescheduled." IF "276 Customer orders are late and have to be rescheduled," THEN **"301 Throughput is decreasing."** IF "278 Unplanned overtime at various WCs must be scheduled to make the missing parts," THEN "279 Bottlenecks seem to wander among WCs." Many manufacturers complain about this effect but do not know that their own policies, procedures, and actions create the "myth" of the wandering bottleneck. In fact, many managers believe that they cannot implement constraints management because their constraint keeps moving. Generally, policies and procedures create the case of the wandering bottleneck. IF "278 Unplanned overtime at various WCs must be scheduled to make the missing parts," THEN "293 Due date performance is decreasing." IF "278 Unplanned overtime at various WCs must be scheduled to make the missing parts," THEN "295 Lead time is increasing." IF "293 Due date performance is decreasing," THEN **"301 Throughput is decreasing."** IF "295 Lead time is increasing," THEN **"301 Throughput is decreasing."**

Is this your reality—operating expense is increasing, inventory is increasing, and throughput is decreasing? Is due date performance decreasing? Are your lead times increasing? Is overtime everywhere? Are customer orders late? Are you constantly rescheduling orders? Are efficiencies and utilizations reduced? To fix these seemingly unrelated "problems," the following core problems must be addressed:

- Using cost accounting to determine the products the organization should produce and sell
- Using a traditional operations performance measurement system that focuses on reducing product costs (uses efficiencies and utilization everywhere as measures)
- Using a traditional scheduling and control system (like MRP)

V-Plant Current Reality Tree

The characteristics of many V-plants are:

- Some WC operations cause a common material to diverge into a specific product. (5/241)
- Setup times are dependent and sometimes long. (5/223)
- All products generally follow the same routing. (5/221)
- The equipment is generally capital intensive and specialized. (5/222)
- The number of products is large compared to the number of raw materials. (5/220)

Let's now examine the CRT for a V-plant provided in Figure 10.5. Do these characteristics fit your plant environment? Is your plant a V-plant? Let's read the bottom entries: "40 First-line supervisors are measured based on resource utilization." "222 In a V-plant, the equipment is generally capital intensive & specialized." "223 In a V-plant, setup times are dependent & sometimes long." "48 First-line supervisors are measured based on direct labor efficiency." IF "40 First-line supervisors are measured based on resource utilization" AND "222 In a V-plant, the equipment is generally capital intensive & specialized" AND "223 In a V-plant, setup times are dependent & sometimes long," THEN "58a First-line supervisors combine batches to eliminate setups at their WC." (Entity 58 was split into 58a and 58b to show separate effects.) IF "222 In a V-plant, the equipment is generally capital intensive & specialized" AND "223 In a V-plant, setup times are dependent & sometimes long" AND "48 First-line supervisors are measured based on direct labor efficiency," THEN "58b First-line supervisors sequence

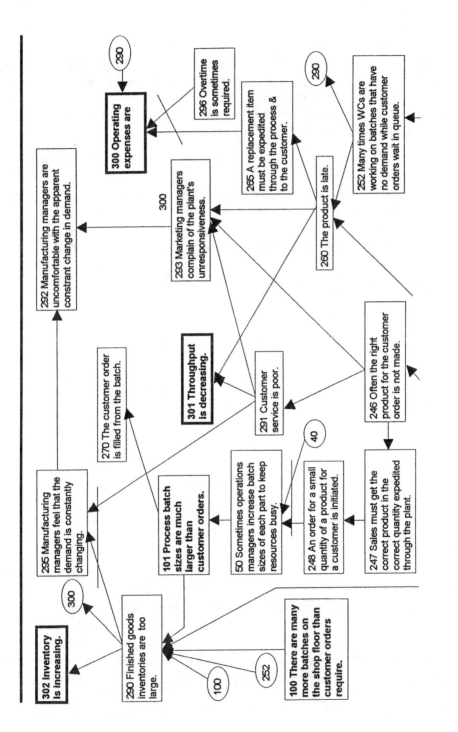

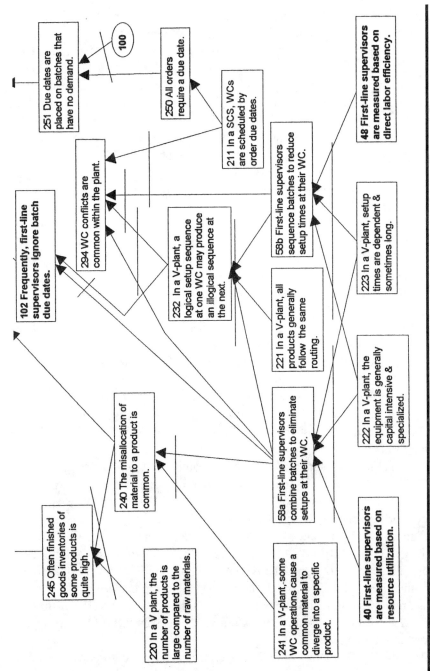

Figure 10.5 Current Reality Tree for a Traditional Scheduling System and Measures in a V-Plant

batches to reduce setup times at their WC." IF "58a First-line supervisors combine batches to eliminate setups at their WC" AND "221 In a V-plant, all products generally follow the same routing" AND "58b First-line supervisors sequence batches to reduce setup times at their WC," THEN "232 In a V-plant, a logical setup sequence at one WC may produce an illogical sequence at the next." IF "58a First-line supervisors combine batches to eliminate setups at their WC" AND "232 In a V-plant, a logical setup sequence at one WC may produce an illogical sequence at the next," THEN "294 WC conflicts are common within the plant." IF "232 In a V-plant, a logical setup sequence at one WC may produce an illogical sequence at the next" AND "58b First-line supervisors sequence batches to reduce setup times at their WC," THEN "294 WC conflicts are common within the plant." IF "58b First-line supervisors sequence batches to reduce setup times at their WC" AND "211 In a SCS, WCs are scheduled by order due dates," THEN "294 WC conflicts are common within the plant." IF "211 In a SCS, WCs are scheduled by order due dates," THEN "250 All orders require a due date." IF "250 All orders require a due date" AND "100 There are many more batches on the shop floor than customer orders require," THEN "251 Due dates are placed on batches that have no demand." IF "251 Due dates are placed on batches that have no demand," THEN "252 Many times WCs are working on batches that have no demand while customer orders wait in queue." IF "252 Many times WCs are working on batches that have no demand while customers orders wait in queue," THEN "290 Finished goods inventories are too large." IF "252 Many times WCs are working on batches that have no demand while customer orders wait in queue," THEN "260 The product is late."

IF "58a First-line supervisors combine batches to eliminate setups at their WC" AND "232 In a V-plant, a logical setup sequence at one WC may produce an illogical sequence at the next," THEN "102 Frequently, first-line supervisors ignore batch due dates." IF "102 Frequently, first-line supervisors ignore batch due dates," THEN "260 The product is late." IF "260 The product is late," THEN "293 Marketing managers complain of the plant's unresponsiveness." IF "260 The product is late," THEN "265 A replacement item must be expedited through the process & to the customer." IF "265 A replacement item must be expedited through the process & to the customer" AND "296 Overtime is sometimes required," THEN **"300 Operating expenses are increasing."**

IF "241 In a V-plant, some WC operations cause a common material to diverge into a specific product" AND "58a First-line supervisors combine batches to eliminate setups at their WC," THEN "240 The misallocation of material to a product is common." IF "240 The misallocation of material to a product is common" AND "220 In a V-plant, the number of products is large compared to the number of raw materials," THEN "245 Often finished goods inventories of some products is quite high." IF "245 Often finished goods inventories of some products is quite high," THEN "290 Finished goods inventories are too large." IF "100 There are many more batches on the shop floor than customer orders require," THEN "290 Finished goods inventories are too large." IF "240 The misallocation of material to a product is common," THEN "246 Often the right product for the customer order is not made." IF "246 Often the right product for the customer order is not made," THEN "293 Marketing managers complain of the plant's unresponsiveness." IF "246 Often the right product for the customer order is not made," THEN "291 Customer service is poor." IF "291 Customer service is poor," THEN "293 Marketing managers complain of the plant's unresponsiveness." IF "293 Marketing managers complain of the plant's unresponsiveness," THEN "292 Manufacturing managers are uncomfortable with the apparent constant change in demand." IF "291 Customer service is poor," THEN **"301 Throughput is decreasing."**

IF "246 Often the right product for the customer is not made," THEN "247 Sales must get the correct product in the correct quantity expedited through the plant." IF "247 Sales must get the correct product in the correct quantity expedited through the plant," THEN "248 An order for a small quantity of a product for a customer is initiated." IF "248 An order for a small quantity of a product for a customer is initiated" AND "40 First-line supervisors are measured based on resource utilization," THEN "50 Sometimes operations managers increase batch sizes of each part to keep resources busy." IF "50 Sometimes operations managers increase batch sizes of each part to keep resources busy," THEN "101 Process batch sizes are much larger than customer orders." IF "101 Process batch sizes are much larger than customer orders," THEN "270 The customer order is filled from the batch." IF "101 Process batch sizes are much larger than customer orders," THEN "290 Finished goods inventories are too large." IF "290 Finished goods inventories are too large" AND "291 Customer service is poor," THEN "295 Manufacturing managers feel that the demand is constantly changing." IF "295 Manufacturing managers feel that the demand is constantly changing," THEN "292 Manufacturing managers are uncomfortable with the apparent constant change in demand." IF "290 Finished goods inventories are too large," THEN "300 Operating expenses are increasing." IF "290

Finished goods inventories are too large," THEN **"302 Inventory is increasing."**

Is this your reality—operating expense is increasing, inventory is increasing, and throughput is decreasing? Is due date performance decreasing and lead time increasing? Is overtime everywhere? Are customer orders late and constantly being rescheduled? Are efficiencies and utilizations reduced? To fix these seemingly unrelated "problems," the following core problems must be addressed:

- Using cost accounting to determine the products the organization should produce and sell
- Using a traditional operations performance measurement system that focuses on reducing product costs (uses efficiencies and utilization everywhere as measures)
- Using a traditional scheduling and control system (like MRP)

T-Plant Current Reality Tree

The characteristics of a T-plant are:

- Fabrication and assembly are treated as separate plants (6/268)
- Several common manufactured and/or purchased components are assembled into the product. (6/220)
- Components are common to many different products. (6/221)
- Due date performance is poor (30 to 40% early and 30 to 40% late). (6/267)

The current reality tree for a T-plant is provided in Figure 10.6. The entities at the bottom of the diagram are: "236 Customers want short lead times." "453 Total lead time is long." "211 In a SCS, fabrication is based on making batches to forecast." "101 Batch sizes are much larger than customer orders." IF "236 Customers want short lead times" AND "453 Total lead time is long," THEN "268 In a T-plant, fabrication & assembly are treated as separate plants." IF "236 Customers want short lead times" AND "268 In a T-plant, fabrication & assembly are treated as separate plants," THEN "233 Products are assembled based on actual customer orders (assemble to order)." IF "210 In a SCS, assembly is scheduled initially based on customer due date," THEN "233 Products are assembled based on actual customer orders (assemble to order)." IF "268 In a T-plant, fabrication & assembly are treated as separate plants," THEN "234 Parts are fabricated & stocked based on forecasts & lot sizing (make to stock)." IF "211 In a SCS, fabrication is based on making batches to forecast" AND "101 Batch sizes are much larger than customer orders," THEN "234 Parts are fabricated & stocked based on forecasts & lot sizing (make to stock)." IF "233 Products are assembled based on actual customer orders (assemble to order)" AND "234 Parts are fabricated & stocked based on forecasts & lot sizing," THEN "201 Sometimes parts are missing when needed." IF "201 Sometimes parts are missing when needed" AND "202 These parts are common parts & assemblies," THEN "205 Cannibalizing of common parts, assemblies, & products occurs." IF "201 Sometimes parts are missing when needed" AND "202 These parts are common parts & assemblies," THEN "206 Stealing of common parts, assemblies, & products occurs." IF "205 Cannibalizing of common parts, assemblies, & products occurs," THEN "216 Common parts & assemblies are missing." IF "206 Stealing of common parts, assemblies, & products occurs," THEN "216 Common parts & assemblies are missing." IF "216 Common parts & assemblies are missing," THEN "217 A customer order is short a component." IF "217 A customer order is short a component," THEN "219 Setups are broken to expedite missing parts." IF 219 Setups are broken to expedite missing parts," THEN "269 Utilization & efficiency are unsatisfactory." IF "269 Utilization & efficiency are unsatisfactory," THEN "58 Frequently, first-line supervisors combine & sequence batches to save on setups." IF "219 Setups are broken to expedite missing parts," THEN "102 Frequently, first-line supervisors ignore batch due dates." IF "217 A customer order is short a component," THEN "218 The customer order is shipped late." IF "218 The customer order is shipped late" AND 271 Due date performance is very important," THEN "222 The customer is upset & may go to a competitor." IF "222 The customer is upset & may go to a competitor," THEN "302 Throughput is decreasing." IF "201 Sometimes parts are missing when needed," THEN "210 Lot sizes are increased to prevent parts shortages." IF "210 Lot sizes are increased to prevent parts shortages," THEN "212 More WIP is on the shop floor." IF "212 More WIP is on the shop floor," THEN "214 Fabrication lead times are increasing." IF "214 Fabrication lead times are increasing" AND "218 A customer order is short a component," THEN "200 Overtime is required to replace missing item." IF "200 Overtime is required to replace missing item," THEN "300 Operating expenses are increasing." IF "200 Overtime is required to replace missing item," THEN "232 Assembly lead times are increased." IF "214 Fabrication lead times are increasing," THEN "242 Component inventories are increasing." IF "214 Fabrication lead times are increasing," THEN "234 Parts are fabricated & stocked based on forecasts & lot sizing (make to stock)." IF

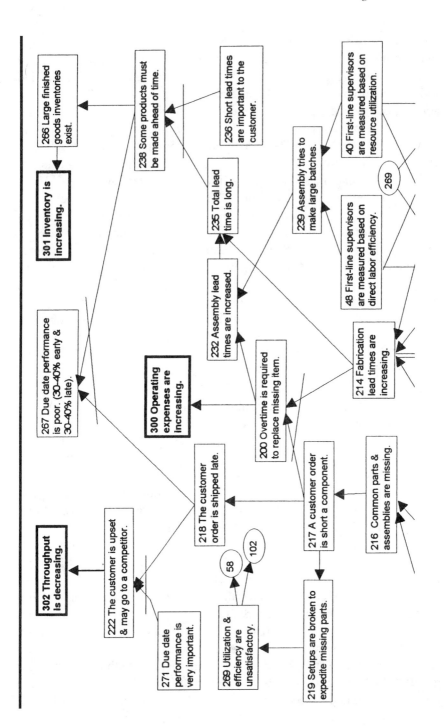

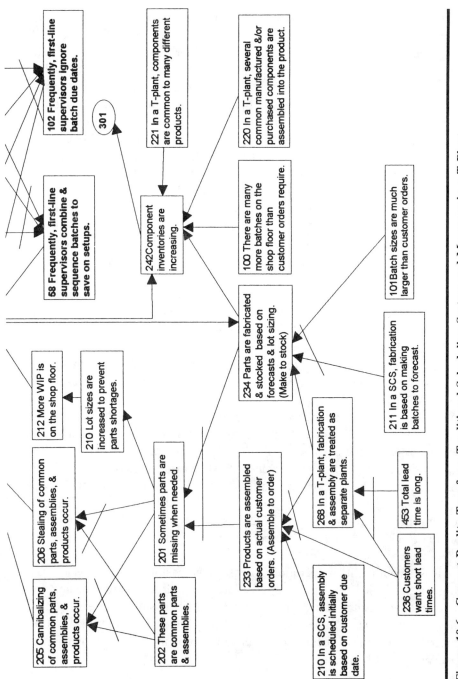

Figure 10.6 Current Reality Tree for a Traditional Scheduling System and Measures in a T-Plant

"234 Parts are fabricated & stocked based on forecasts & lot sizing (make to stock)," THEN "242 Component inventories are increasing." IF "100 There are many more batches on the shop floor than customer orders require," THEN "242 Component inventories are increasing." IF "220 In a T-plant, several common manufactured &/or purchased components are assembled into the product," THEN "242 Component inventories are increasing." IF "221 In a T-plant, components are common to many different products," THEN "242 Component inventories are increasing." IF "242 Component inventories are increasing," THEN "301 Inventory is increasing." IF "48 First-line supervisors are measured based on direct labor efficiency," THEN "58 Frequently, first-line supervisors combine & sequence batches to save on setups." IF "48 First-line supervisors are measured based on direct labor efficiency," THEN "102 Frequently, first-line supervisors ignore batch due dates." IF "40 First-line supervisors are measured based on resource utilization," THEN "58 Frequently, first-line supervisors combine & sequence batches to save on setups." IF "40 First-line supervisors are measured based on resource utilization," THEN "102 Frequently, first-line supervisors ignore batch due dates." IF "40 First-line supervisors are measured based on resource utilization," THEN "239 Assembly tries to make large batches." IF "48 First-line supervisors are measured based on direct labor efficiency," THEN "239 Assembly tries to make large batches." IF "239 Assembly tries to make large batches," THEN "232 Assembly lead times are increased." IF "232 Assembly lead times are increased," THEN "235 Total lead time is long." IF "214 Fabrication lead times are increasing," THEN "235 Total lead time is long." IF "235 Total lead time is long" AND "236 Short lead times are important to the customer," THEN "238 Some products must be made ahead of time." IF "238 Some products must be made ahead of time," THEN "266 Large finished goods inventories exist." IF "266 Large finished goods inventories exist," THEN "301 Inventory is increasing." IF "218 The customer order is shipped late" AND "238 Some products must be made ahead of time," THEN "267 Due date performance is poor (30-40% early & 30-40% late).

Is this your reality—operating expense is increasing, inventory is increasing, and throughput is decreasing? Is due date performance decreasing and lead time increasing? Is overtime everywhere? Are customer orders late? Do they constantly have to be rescheduled? Are efficiencies and utilization reduced? To fix these seemingly unrelated "problems," the following core problems must be addressed:

- Using cost accounting to determine the products the organization should produce and sell
- Using a traditional operations performance measurement system that focuses on reducing product costs (uses efficiencies and utilization everywhere as measures)
- Using a traditional scheduling and control system (like MRP)

THE TEMPORARY SOLUTION TO YOUR CORE PROBLEMS

Notice that we keep coming back to the same core problems. Solving these core problems is certainly fundamental to temporarily moving your organization to a process of continuous improvement. There are tools to develop an effective production plan based on using your resource capabilities as a production system linked to customers' needs and to organizational profit. In Chapters 3, 4, and 5, we discussed the five focusing steps, the drum-buffer-rope scheduling method, and V-A-T analysis, respectively. In this chapter, we discussed the product mix problem. Are you getting the most out of your resources? Are you using cost accounting to determine product costs and this, in turn, to determine your desired product mix, the addition of new products, and the elimination of old products? Is your product cost used to set product price and, in turn, are sales commissions based on product price?

Implementing a constraint management performance measurement system and using the five focusing steps are basic steps toward getting the most from your existing resource base. These steps provide the basis for developing a logical production plan. This plan will have the most impact on the organizational profit given the organization's resource base. This product mix should be the basis for decisions regarding product pricing, sales commissions, marketing opportunities, future growth, etc. This plan allows the organization to get the most from its resources. It should provide the basis for all local performance measures.

Implement drum-buffer-rope to schedule the constraint and schedule the control points (material release or gating work centers, divergent work centers, convergent work centers, shipping work center, and the associated buffers) based on the constraint schedule. Only a few work centers are provided detailed schedules, in contrast to MRP where every work center is given a detailed schedule of activities. Remember, in drum-buffer rope, workers are instructed, "Work if you have work. If you don't have work, don't worry; it will be coming to you." This idle time is your protective capacity. Do not fill it up to keep workers busy. Of course, the schedule is linked to the production plan and is also linked to the control function. The buffers provide the control of the flow of products. Periodically check to determine if the items that are supposed to be in the buffer are actually there yet. If a work order is in region 1 of the schedule and is not at the control point, then walk back down the work order's routing and determine its location. Expedite it to the control point. This is true "management by exception." Here, exceptions impact the bottom line of the organization. Use buffer management to identify improvement opportunities in the production system.

You have now taken the initial steps required for continuous improvement. You have straightened out your production planning process, your scheduling process, your control process, and your production system improvement process. But what happens when the constraint moves from the plant (the production function) to another function such as marketing, sales, engineering, purchasing, etc.? What should you do? Will you do what so many other companies have done? Will you lay off your workers, staff, and managers? Is this the ultimate result of organizational improvement—people laid off? Reengineering in the name of organizational improvement?

We don't think so! Review Chapter 2 again and review your organization's situation. Look again at Figure 2.2. Where are you in respect to this framework? Where does your company want to go? Next, read Chapter 11 and learn to construct a CRT. You control your company's destiny. The constraint will move when you take the steps outlined in this handbook. The question for your company (for its resource capabilities and market opportunities) is where the constraint should be located. This question must be answered if an organization is to become and remain a world-class organization and is to attain and remain on a path of continuous organizational improvement. The thinking tools are essential for proper management of any organization.

A closing note on performance measurement and the logistics paradigms is in order. Any performance measurement system must link market potential and opportunities to the incremental cost associated with the resources providing the goods or services to meet the potential and opportunities. Both resource capabilities and customers' needs must be continually evaluated to assess the bottom-line impact of market opportunities. This is the only way a process of continuous improvement can be sustained—by continually growing new market opportunities and satisfying loyal customers. You satisfy loyal customers by continually meeting or exceeding their needs for products and services with respect to variety, price, quality, lead time, due date performance, new product introduction, responsiveness, etc. You must continually provide a product or service that meets or exceeds customer expectations when purchased and during its life. This approach guarantees future business. The Thinking Process again can help you solve customer problems.

In this chapter, the underlying rationale for the constraints management performance measurement system was presented. Note how the CRTs all lead to the same conclusion—traditional measurement systems cannot support managerial decisions we now need to make. The creation and use of throughput, inventory, and operating expense as the appropriate organizational measures do lead to correct managerial decisions.

REFERENCES

1. Goldratt, E.M., "What is the Theory of Constraints?" *APICS The Performance Advantage,* 3(6), 18-20, 1993.
2. Cox, J.F., J.H. Blackstone, and M.S. Spencer, *APICS Dictionary,* Falls Church, VA: American Production and Inventory Control Society, 1995.
3. Goldratt, E.M., *It's Not Luck,* Croton-on-Hudson, NY: North River Press, 1994.
4. Lockamy, A. and J.F. Cox, *Reengineering Performance Measurements: How to Align Systems to Improve Processes, Products, and Profits,* Burr Ridge, IL: Irwin Professional Publishing, 1994.
5. Goldratt, E.M. and R.E. Fox, *The Race,* Croton-on-Hudson, NY: North River Press, 1986.
6. Goldratt, E.M., *The Haystack Syndrome: Sifting Information Out of the Data Ocean,* Croton-on-Hudson, NY: North River Press, 1990.
7. Goldratt, E.M. and J. Cox, *The Goal: A Process of Continuous Improvement,* 2nd edition. Croton-on-Hudson, NY: North River Press, 1994.

GOLDRATT'S THINKING PROCESSES

INTRODUCTION

What are the Goldratt thinking processes? They are a set of stand-alone or linked logic tools based on causal relationships. The tools are useful in identifying the core problem in a personal, organizational, or any situation; determining and testing a win–win solution prior to implementation; and determining the obstacles to implementation and how to address them. Each of the six tools has a wide array of uses too numerous to list and discuss in this last chapter; therefore, detailed instructions are provided for constructing the current reality tree, the evaporating cloud, and the negative branch reservation tools. Chapter 10 provided specific applications that relate to measures and A-, V-, and T-plants as they provide linkages from the traditional performance measure system and the traditional scheduling system to the specific problems. The evaporating cloud will be used to illustrate the conflict managers face in implementing a true process of continuous improvement in an organization.

As indicated in the introduction in Chapter 1, many firms have successfully implemented various applications of Theory of Constraints (TOC) but are quite reluctant to discuss the details or admit to using TOC. Some see it as such a competitive edge that they do not want rivals to know how they are achieving their results. Some have integrated it into their existing just-in-time and/or total quality management programs not wanting employees to think another new buzzword is being implemented. Surprisingly enough, in a recent research effort, an audience of over 200 TOC implementors were asked to share their successes and failures with the authors. Goldratt made a plea to share their experience publicly. Over 30 professionals provided their business cards. After two letters of inquiry and offers to these professionals from the authors to take their notes and write a draft, only one contribution has been received. In later personal conversations with some, they admitted that upper management had rejected the idea of sharing their successes with the public. Management was afraid that competitors would use the information to catch up with them, and they were unsure of how to respond to the renewed intense competition. They were pleased to manage where they felt they were clearly superior to rivals.

WHAT IS THE ROLE OF A MANAGER IN TODAY'S COMPETITIVE ENVIRONMENT?

Goldratt[1] states that the role of a manager is to determine:

- What to change
- What to change to
- How to successfully cause the change

In today's competitive environment, seemingly hundreds of problems and potential solutions exist. In manufacturing, most organizations have problems meeting customers' increasing demands for more product variety, lower prices, higher quality, shorter lead times, better due date performance, etc. Problems exist with workers, managers, equipment, facilities, etc. Major problems exist with vendors: selection, prices, quality, lead times, due date performance, etc. Hundreds of problems crop up when the problems and functions are viewed as independent of each other. The real question for a manager is "What is(are) the problem(s) that must be addressed to increase organizational productivity?" Where are the significant organizational gains to be made? Managers tend to firefight, to address the loudest screams, to focus where an urgent (but probably unimportant) need arises. This problem is seldom the area where true organizational progress can be made. But managers expend enormous energy fighting fires. Many view themselves as heroes when they resolve such problems. These daily problems are generally the symptoms of deeper rooted problems—problems that never get addressed, problems that most managers do not even recognize as problems. Until solved, these problems will continually cause the day-to-day need for firefighting. **Correct identification of the core problem(s) must be a primary responsibility of a manager. As discussed in Chapter 10, the current reality tree provides managers with this skill.**

PROCEDURE FOR CONSTRUCTING A CURRENT REALITY TREE

The procedure for constructing a current reality tree[2]* is as follows:

1. List between five and ten problems (called undesirable effects [UDEs]) related to the situation.
2. Test each UDE for clarity. Is the UDE a clear and concise statement? This test is called the *clarity reservation.*
3. Search for a causal relationship between any two of the UDEs.
4. Determine which UDE is the cause and which is the effect. Read as "IF *cause,* THEN *effect.*" This test is called the *causality reservation.* Occasionally the cause and effect may be reversed. Check by using the following statement: "*Effect* BECAUSE *cause.*"
5. Continue the process of connecting the UDEs using the IF-THEN logic until all UDEs are connected.
6. Often, the causality is strong to the person feeling the problem but does not seem to exist to others. In these instances, "clarity" is the problem. Use the *clarity reservation* to eliminate the problem. Generally, entities between the cause and the effect are missing. The current relationship is stated as "IF *cause,* THEN *effect.*" The correct structure in its simplest form may be "IF *cause,* THEN (*missing effect*). (The missing effect becomes the missing cause at the next higher level: IF (*missing cause*), THEN *effect.*"
7. Sometimes the cause by itself may not seem to be enough to create the effect. These cases are tested with the *cause insufficiency reservation* and are improved by reading: "IF *cause* and _____, THEN *effect.*" What is the missing *dependent* statement that completes the logical relationship? Add it to your diagram using the *and connector* (represented graphically by a horizontal line across both connecting arrows). The AND in this relationship is called a *conceptual and,* which means that both entities connected with the *and connector* have to be present for the effect to exist.
8. Sometimes the effect is caused by many *independent* causes. The causal relationships are strengthened by the *additional cause reservation.* The problem to be addressed is: "How many of the causes are important enough to address?" One, two, or sometimes three

---

* The guidelines are modifications of the procedure outlined in Chapters 12 through 16 of Goldratt, Eliyahu M., *It's Not Luck.*

causes frequently result in creating about 80% of the effect. Generally, eliminating these few causes is enough of a reduction so that the remaining effect becomes minor. Therefore, it is not necessary to have an exhaustive list of causes for an effect. The *and connector* for these independent causes is called a *magnitudinal and*, which means that each of the causes must be addressed to eliminate most of the effect. This *magnitudinal and connector* exists where two or more arrow tips enter an entity. No horizontal line is used to connect the arrows.

9. Sometimes an IF-THEN relationship seems logical but the causality is not appropriate in its wording. In these instances, words like "some," "few," "many," "frequently," "sometimes," and other modifiers can make the causality stronger.

10. Numbering of UDEs on the CRT is for ease of locating entities only. An asterisk by a UDE indicates the UDE was provided in the original list of UDEs.

An Application

In Chapter 10, the CRT for a traditional operations performance system was presented and studied. That performance measurement CRT, given in Figure 11.1, will now be used to illustrate the procedure for constructing a CRT.

Step 1. List between five and ten problems (UDEs)

1. Frequently, first-line supervisors ignore due dates.
2. There are more parts on the shop floor than customer orders require.
3. Reducing cost is emphasized as the goal for operations managers.
4. Batch sizes are much larger than customer orders.
5. Large batches are released to keep resources busy.
6. Frequently, first-line supervisors combine and sequence lots to save time.

Step 2. Test each UDE for clarity

Is the UDE a clear and concise statement? This test is called the *clarity reservation*. The original UDEs were edited to improve clarity.

1. Frequently, first-line supervisors ignore batch due dates. (102)
2. There are many more batches on the shop floor than customer orders require. (100)

3. Reducing product cost is emphasized as the goal for operations managers. (20)
4. Process batch sizes are much larger than customer orders. (101)
5. Sometimes operations managers release parts early to keep workers busy. (55)
6. Frequently, first-line supervisors combine and sequence batches to save time on setups. (58)

Step 3. Search for a causal relationship between two of the UDEs

Both entities 2 and 5 and entities 4 and 6 seem to be related. Let's check entities 2 and 5 first. Instead of trying to keep track of the specific numbers of UDEs in the diagram, let's just use any number (step 10). Remember, numbering is to assist us in quickly locating the entity on the diagram and nothing more. We will use an asterisk after an entity to identify it as one of the original UDEs.

Step 4. Determine which UDE is the cause and which is the effect

Read as "IF *cause* THEN *effect*." This test is called the *causality reservation*. Occasionally, the cause and effect may be reversed. Check by using the following statement: "*Effect* BECAUSE *cause*." IF "sometimes operations managers release parts early to keep workers busy," THEN "there are many more batches on the shop floor than customer orders require." This relationship seems logical and causal except for the modifiers "sometimes" and "many more." This point is discussed in step 9. Let's not change these modifiers now. We will revisit this issue later, as other items may be causally related to the effect. This part of the CRT diagram is provided below:

100 There are many more batches on the shop floor than customer orders require.*

↑

55 Sometimes operations managers release parts early to keep workers busy.*

Let's check the second causal relationship:

101 Process batch sizes are much larger than customer orders.*

↑

58 Frequently, first-line supervisors combine and sequence batches to save time on setups.*

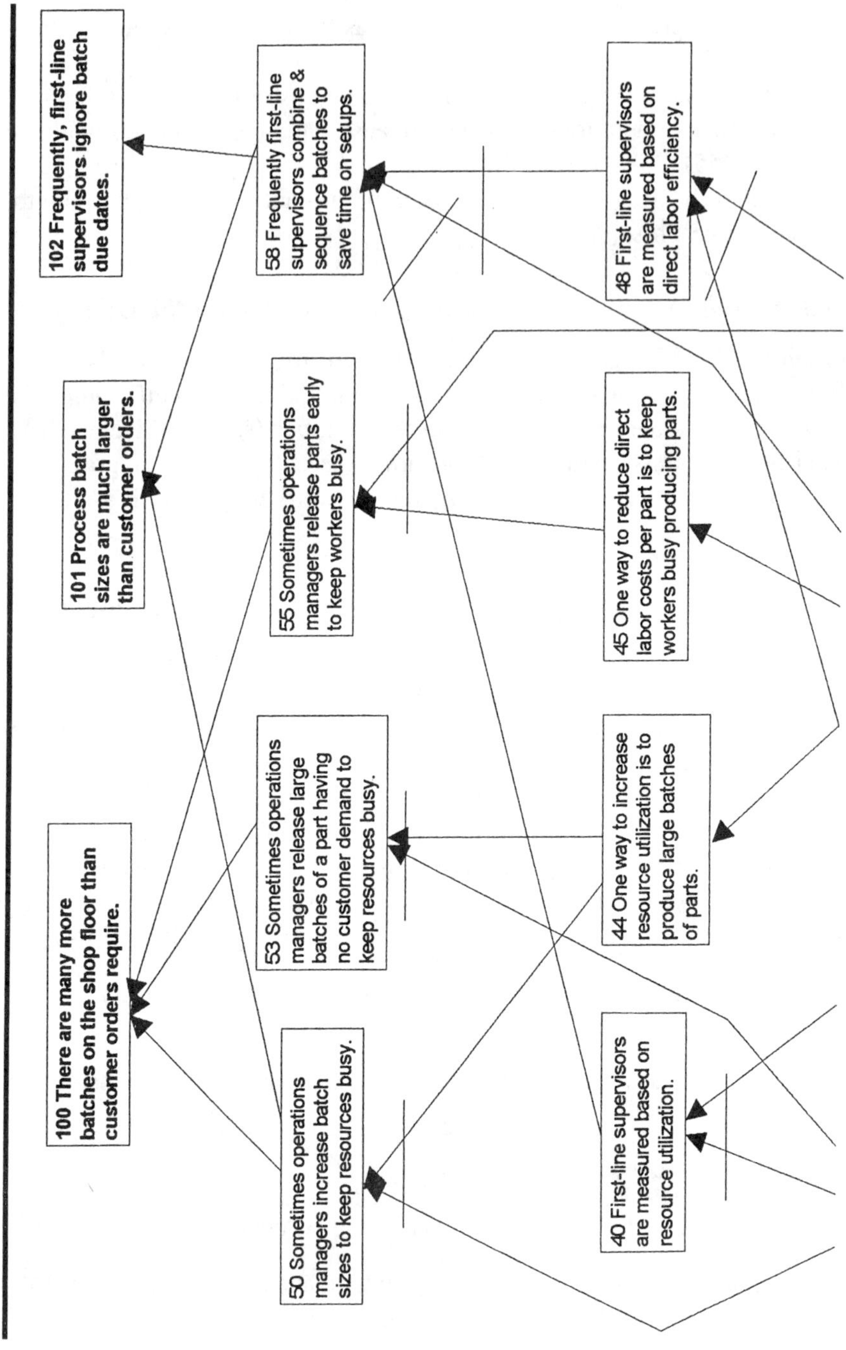

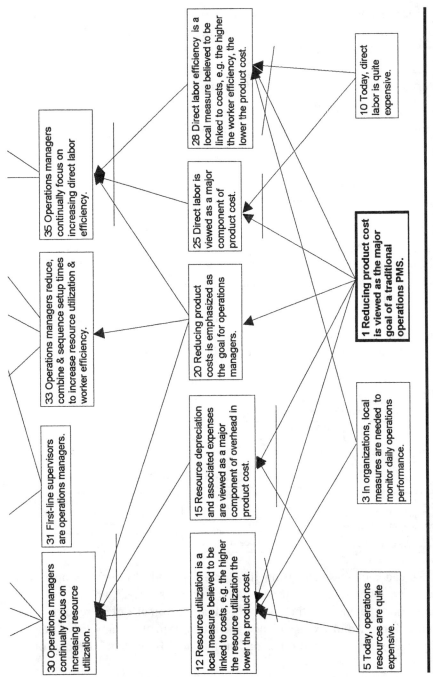

Figure 11.1 Current Reality Tree of the Traditional Operations Performance Measurement System in a Manufacturer

This statement also seems logical and passes the test of reading as "process batch sizes are much larger than customer orders" BECAUSE "frequently, first-line supervisors combine and sequence batches to save time on setups."

Step 5. Continue the process of connecting the UDEs until all are connected

Entity 6 also seems to cause entity 1. This relationship is given below:

102 Frequently, first-line supervisors ignore batch due dates.*

↑

58 Frequently, first-line supervisors combine and sequence batches to save time on setups.*

The remaining rules are not necessarily applied in the number sequence presented. They are rules to clarify the causal relationships by modifying the wording or structure of the entities. An example of applying step 8 is provided. There seem to be additional causes for UDE 100: "There are many more batches on the shop floor than customer orders require." Why are there more batches? "100 There are many more batches on the shop floor than customers' orders require" BECAUSE "50 Sometimes operations managers increase batch sizes to keep resources busy."

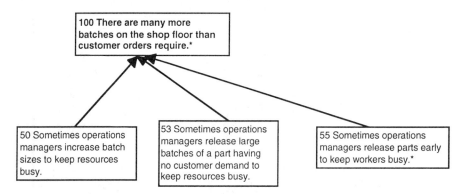

Reread the statement using IF-THEN logic: IF "sometimes operations managers increase batch sizes to keep resources busy," THEN "there are many more batches on the shop floor than customer orders require." This sounds logical and exists in most facilities. Are there any other reasons why "there are many more batches on the shop floor than customer orders require"? Well, "sometimes operations managers release large batches of a part having no customer demand to keep resources busy." There is no customer demand for the part, but the manager wants to keep resources busy and releases part orders that have no demand yet. Are there any other major reasons why "there are many more batches on the shop floor than customer orders require"? No, we have identified three major causes which account for most of the effect. These relationships illustrate the structure of the *magnitudinal and.* Let's read what we now have:

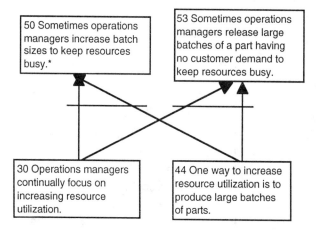

Now let's delve deeper into our understanding of the manufacturing environment. Entities 50 and 53 seem to be related. What causes them? "Sometimes operations managers increase batch sizes to keep resources busy" BECAUSE "one way to increase resource utilization is to produce large batches of parts" AND "operations managers continually focus on increasing resource utilization." Lets read the statements in the *conceptual and* format of IF-AND-THEN provided in step 7. IF "30 Operations managers continually focus on increasing resource utilization" AND "44 One way to increase resource utilization is to produce large batches of parts," THEN "50 Sometimes operations managers increase batch sizes to keep resources busy." IF "30 Operations managers continually focus on increasing resource utilization" AND "44 One way to increase resource utilization is to produce large batches of parts," THEN "53 Sometimes operations managers release large batch sizes of a part having no customer demand to keep resources busy." These *conditional and* relationships are as follows:

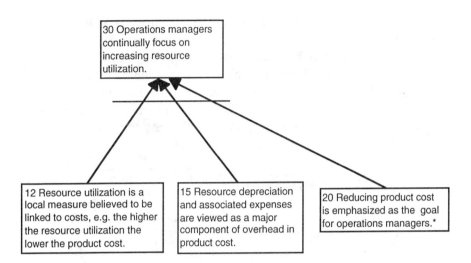

We still have not connected all of the original UDEs, nor does it seem that we understand the situation deeply enough to identify a core problem. We have UDE 20: "Reducing product cost is emphasized as the goal for operations managers." This seems to be related to UDE: "Operations managers continually focus on increasing resource utilization." But what is missing? IF "20 Reducing product cost is emphasized as the goal for operations managers" AND___ THEN "30 Operations managers continually focus on increasing resource utilization." How do we fill in the blank—what is the logical connection? IF "20 Reducing product cost is emphasized as the goal for operations managers" AND "15 Resource depreciation and associated expenses are viewed as a major component of overhead in product cost" AND "12 Resource utilization is a local measure believed to be linked to costs, e.g., the higher the resource utilization, the lower the product cost," THEN "30 Operations managers continually focus on increasing resource utilization." This logic is sound.

What about entity 12? Why does "12 Resource utilization is a local measure believed to be linked to costs, e.g., the higher the resource utilization, the lower the product cost" exist? It exists because "5 Today, operations resources are expensive" AND "3 In organizations, local measures are needed to monitor daily operations performance" AND "1 Reducing product cost is viewed as the major goal of a traditional operations PMS."

Why does entity 15 exist? "15 Resource depreciation and associated expenses are viewed as a major component of overhead in product cost" BECAUSE "5 Today, operations resources are quite expensive" AND "1

Reducing product cost is viewed as the major goal of a traditional operations PMS." Of course, If "1 Reducing product cost is viewed as the major goal of a traditional operations PMS," THEN "20 Reducing product costs is emphasized as the goal for operations managers." These relationships are shown below.

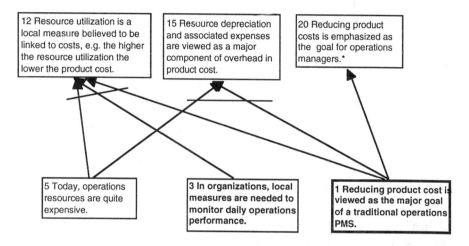

| 12 Resource utilization is a local measure believed to be linked to costs, e.g. the higher the resource utilization the lower the product cost. | 15 Resource depreciation and associated expenses are viewed as a major component of overhead in product cost. | 20 Reducing product costs is emphasized as the goal for operations managers.* |

| 5 Today, operations resources are quite expensive. | 3 In organizations, local measures are needed to monitor daily operations performance. | 1 Reducing product cost is viewed as the major goal of a traditional operations PMS. |

Today, direct labor is quite expensive also. Hundreds of plants have moved overseas to reduce labor costs. What is the impact on measures and ultimately on policies within the plant? IF "1 Reducing product cost is viewed as the major goal of a traditional operations PMS" AND "10 Today, direct labor is quite expensive," THEN "25 Direct labor is viewed as a major component of product cost." Additionally, IF "1 Reducing product cost is viewed as the major goal of a traditional operations PMS" AND "10 Today, direct labor is quite expensive" AND "3 In organizations, local measures are needed to monitor daily operations performance," THEN "28 Direct labor efficiency is a local measure believed to be linked to costs, e.g., the higher the worker efficiency, the lower the product cost." The logical connections continue upward to link the efficiency measure to several of the UDEs initially provided. This development is left as a test for the reader.

The completed CRT is given in Figure 11.1. What is the core problem of this CRT? Entity 1—"Reducing product cost is viewed as the major goal of a traditional operations PMS"—seems to be connected to most of the UDEs. (Remember, by definition, a core problem should be connected to at least 70% of the UDEs.) In fact, every entity in the diagram is causally connected to the core problem. If this problem can be addressed, then almost all of the

UDEs disappear. Our task in this exercise was not to solve the performance measurement problem of a traditional manufacturer but to illustrate the procedure for constructing a CRT. These construction rules apply to developing the most complex CRT diagrams—no more rules are required. Actual construction of CRTs is a very difficult and time-consuming task.* However, it is a skill all business professionals need in today's complex business environment. Causality diagramming will be as essential to business professionals as calculus is essential to engineers.

Causality diagramming provides the mechanism for:

- Identifying the impact of policies, procedures, and actions in one functional area on other organizational areas and ultimately on the organization's goal
- Communicating clearly and concisely the causality of these policies, procedures, and actions
- Identifying the core problem in a given situation
- Building teamwork by moving from the traditional finger-pointing environment to "you and me" against the problem

In Chapter 10, CRTs were provided for A-, V-, and T-plant structures (Figures 10.4, 10.5, and 10.6, respectively) to help explain the causality that exists in many organizations; the construction procedure will not be discussed further. In all instances, diagrams are read from the bottom to the top using IF-THEN and IF-AND-THEN logic.

THE EVAPORATING CLOUD—A CREATIVE PROCESS

Once a core problem has been identified, the manager must search for and identify a win–win solution. Four outcomes are possible in the negotiation of a solution to a problem, as shown in Figure 11.2. Whichever side is the strongest can impose its solution on the other side. If side 1 imposes his or her solution on side 2, then this is a win–lose solution—a win for side 1 and a loss for side 2. If side 2 imposes his or her her solution on side 1, then this is a lose–win solution—a loss for side 1 and a win for side 2. If both sides are equally powerful, then both sides can force a compromise solution where neither side is satisfied with the solution; this solution is a lose–lose solution—a loss for side 1 and a loss for side 2.

---

* Participants in a two-week Jonah course spend four days constructing a CRT of their "business" problems.

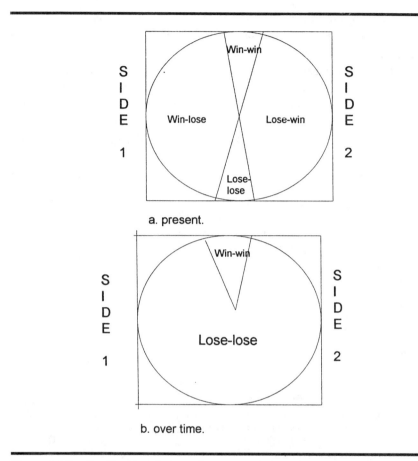

a. present.

b. over time.

Figure 11.2   Conflict Resolution—Present and Long-Term Perspective

Both side 1 and side 2 can search diligently for a win–win solution—a win for side 1 and a win for side 2. These win–win solutions are usually very difficult to identify. The only long-lasting solution, however, is a win–win solution. If both sides do not strive for and implement a win–win solution, then the implemented solution over time will degenerate into a lose–lose solution (see Figure 11.2b).

What tools are available to help you as a manager identify and test a win–win solution? Goldratt's evaporating cloud technique is an excellent tool for precisely defining a problem. Once the problem is defined, you can surface the assumptions related to the problem statement, its requirements, and the conflict. In examining these assumptions, you are able to challenge them to determine if finding an action that will break the assumption can offer a potential avenue for solving the problem.

The general format of an evaporating cloud is provided in Figure 11.3. Objective A is the common goal that is to be achieved by both sides. The goal is the reversal or elimination of the core problem from the CRT. It also may be any other objective that you would like to achieve but are blocked from achieving by some conflict. The requirements, B and C, are what each

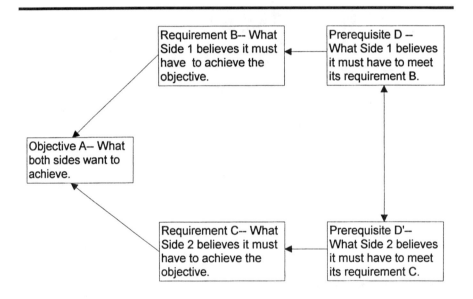

Surfacing the assumptions and injections.

In order to do A, "I" must do B because assumption AB.

In order to do B, "I" must do D because assumption BD.
In order to do A, "I" must do C because assumption AC.
In order to do C, "I" must do D' because assumption DD'.
On one hand I must do D, but on the other I must do D'. Why is there a conflict?
Is there an injection that invalidates the assumption?

Figure 11.3   Structure of Goldratt's Evaporating Could (Modified from Goldratt, Eliyahu M., *What Is This Thing Called Theory of Constraints?* **Croton-on-Hudson, NY: North River Press, 1990 and** *It's Not Luck,* **Croton-on-Hudson, NY: North River Press, 1995)**

side, respectively, or the problem solver believes are the underlying foundations that are required to achieve the objective. Prerequisite D is believed necessary to achieve requirement B and prerequisite D′ is believed necessary to achieve requirement C. The conflict is that both D and D′ cannot exist simultaneously—more of one means less of the other, having one may mean not having the other, etc. To surface the assumptions related to the conflict, the following statement should be completed: In order to have (the tip of the arrow), we must have (the tail of the arrow) because assumption (tip-tail of the arrow). This framework is useful in surfacing assumptions for AB, BD, AC, and CD′. To surface the assumption for D-D′, the following statement is useful: On one hand, we must have D, but on the other hand, we must also have D′. What (*assumption*) prevents us from having both D and D′?

Let's look at a specific application of the evaporating cloud to illustrate its usefulness. See Figure 11.4 for Goldratt's dilemma of maintaining a process of continuous improvement. The objective, A, is to "put an organization on a process of continuous improvement." The requirements that must be met to achieve this objective are "B—induce people to improve" and "C—convert local improvements into bottom-line results." In order to (achieve requirement B) "B—induce people to improve," we must (prerequisite D) "D—not lay off the people who contribute to improvement." In order to (achieve requirement C) "C—convert local improvements into bottom-line results," we must (prerequisite D′) "D′—lay off people in the departments that have improved the most." On one hand, "D—we must not lay off people who contribute to improvement," while on the other hand, we must "D′—lay off people in the departments that have improved the most." This cloud sounds like quite a conflict. This conflict has been played out over and over again during the late 1980s and 1990s in numerous organizations. How can it be resolved where the objective is met without a compromise?

Let's surface the assumptions of our dilemma by using the following statements. In order to put an organization on a path of continuous improvement, we must induce people to improve because motivated people seek out improvement opportunities. An injection would be to reward people for identifying and implementing their improvement opportunities. This injection actually strengthens the assumption by objectively providing a strong motivating force for improvement. Let's continue since this injection does not really solve the conflict. In order to induce people to improve, we must not lay off people who contribute to improvement. This is probably why so many business process reengineering (BPR) projects have failed. Most BPR projects have focused on saving money by replacing people with technology.

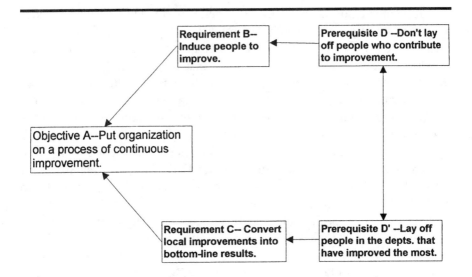

Assumptions and Injections:
AB-Motivated people seek out improvement opportunities.
  -Reward people for contributions to organizational improvement.
BD-People do not want to improve themselves out of a job.
  - Guarantee job security.
AC-Local improvement measures link directly to bottom line results.
  -Use TIOE and the Five Focusing Steps to identify true improvements.
CD-The market is limited.
  - Aim improvements at expanding the market.
DD'-Don't layoff and layoff people.
  -Move ("layoff") from the department where they have improved and move
  them to departments that need improvement as facilitators.

Figure 11.4   Goldratt's Process on Ongoing Improvement Evaporating Cloud (Modified from Goldratt, Eliyahu, "The Theory of Constraints in Industry," 1996 APICS Constraint Management Symposium Proceedings, Detroit, MI. With permission.)

The saving is realized by laying off the people who create the improvement. No wonder few BPR efforts have been successful. Who wants to recommend process improvements that get you laid off? In order to put an organization on a path of continuous improvement, we must convert local improvement into bottom-line improvements because local measures show us where we can save money. An injection might be to implement measures and a process that help us identify where our improvement efforts allow the firm to make more money.

In order to convert local improvement into bottom-line results, we must lay off people who have improved the most because our market is limited. An injection is to identify opportunities that allow the firm to aggressively attack the existing or new markets. On one hand, we cannot lay off people who contribute to improvement, while on the other hand, we must lay off people in the departments that have improved the most. Are they the same people? An injection is that we can shift people around as we expand the market. We can reduce the number of people in the department where improvement created excess personnel and move them to the departments where growth has created a need for personnel. Notice that these solutions all attack the silo structure of most organizations. Organizational improvement is defined as improvement in bottom-line results, which translates into selling the improvement to the market in the form of better products and services.

NEGATIVE BRANCH RESERVATION

The negative branch reservation (NBR) is another useful tool. It is used to test the feasibility and negative effects of taking an action in a specific situation. The NBR is a powerful "stand-alone" logic tool. Goldratt provides the rules for constructing and reading the NBR and an example of its use in *It's Not Luck.*[1] The rules of construction are summarized as follows:

1. Write a brief story line describing the situation.
2. List the positive effects of taking the action to resolve the situation.
3. List the negative effects of taking the action to resolve the situation.
4. Start with the proposed action written at the bottom of the page and use IF-THEN logic to build upward the causal relationships between the action and each negative effect.
5. Examine each statement for logic errors. Check for clarity. Is each statement complete, concise, and logical? Again, check the IF-THEN connection for clarity. Using the IF-THEN linkage, does the statement make sense? This is the *clarity reservation*. Does it represent a causal relationship? IF *cause* THEN *effect*. This is the *causality reservation*. Is the cause sufficient to create the effect? Is something missing from the causal linkage? This is the *cause insufficiency* reservation.

6. Rewrite the branch to eliminate any offensive language. You do not want to create an argument while trying to find a win–win solution.

Application of NBR: Automation or Focus?*

Step 1. Write a brief story line describing the situation

BOB'S BOLT COMPANY STORY LINE—Bill Bell, the plant manager, is approached by his equipment services support supervisor who had recently returned from a trade exhibition where he had observed several AMTs (advanced manufacturing technologies) and had been impressed by their operations under simulation. The supervisor excitedly suggests that the plant should seriously consider installing several of these systems to overcome some specific operational and technical problems that were currently being experienced. Bill had been similarly impressed after an earlier visit to the trade exhibition and understood his supervisor's enthusiasm. However, he had recently read *The Goal*,[4] *It's Not Luck*,[1] and *Reengineering Performance Measurement*[5] and had heard about the TOC Thinking Processes. He knew the problems of poor measures and lack of focus. He wanted to consider several managerial and scheduling options before considering an investment of the size suggested by the supervisor. While contemplating his answer to the supervisor, Bill's secretary hands him a note from the CEO, Bob, which informs him that their major competitor has installed several of the systems the supervisor had suggested. Bob has asked Bill to meet him this afternoon to discuss the immediate purchase of similar systems. Bill tells his supervisor, "We'll discuss this later." Unfortunately, the same response is not suitable for Bob.

FOR MEETING WITH BOB

Step 2. List the positive effects of taking the action to resolve the situation

Positive Effects

1. Bill knows from the literature that these systems have several benefits:
   • More flexibility

---

* This scenerio was provided by a student of the Avraham Y. Goldratt Institute, Management Skills Workshop, University of Georgia, June 1995.

- Ability to change production lot sizes
- Faster time to market

2. Implementation would allow the plant to meet its competitor's initiative.

Step 3. List the negative effects of taking the action to resolve the situation

Negative Effects

1. Serious risk of failure if implemented too hurriedly.
2. We will be less competitive if we spend the money and gain no real benefit (no increase in throughput [T] or reduction in inventory [I] or reduction in operating expense [OE]).
3. Workers are not skilled in operating these systems.
4. Other innovations may be more productive.

Step 4. Start with the proposed action written at the bottom of the page and use IF-THEN logic to build upward the causal relationships between the action and each negative effect

The proposed action is: "We purchase and implement several AMTs." Several of the negative effects are causally linked to each other. We will build them separately using the remaining rules and leave it to the reader to combine the NBRs into one tree. Also, you should be able to identify injections based on reading the previous chapters to eliminate the negative consequences of this proposed action.

The first NBR is in Figure 11.5. IF "10 We are in a highly competitive environment" AND "20 Our competitors have implemented several AMTs," THEN "30 We feel we are falling behind technologically." IF "30 We feel we are falling behind technologically" AND "40 We purchase and implement several AMTs" AND "45 We don't know how to determine the weakest link of the chain," THEN "50 We may choose the wrong AMTs to improve organizational productivity." IF "50 We may choose the wrong AMTs to improve organizational productivity" AND "60 The weakest link determines the strength of the chain," THEN "70 We may increase I and OE without increasing T." IF "70 We may increase I and OE without increasing T," THEN "80 We may be less competitive in our market." Does this sound logical? Will the action of immediately purchasing several AMTs logically

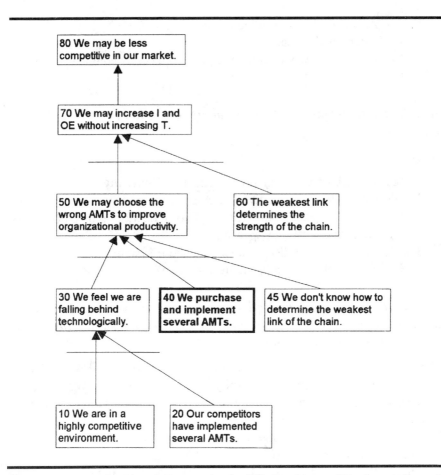

Figure 11.5   Negative Branch Reservation for "We Purchase and Implement Several AMTs"

lead to being "less competitive"? What actions could eliminate this undesirable effect?

In Figure 11.6, the NBR for "We may select an AMT that is not suited to our workers' skills" is constructed. The base is the same. IF "10 We are in a highly competitive environment" AND "20 Our competitors have implemented several AMTs," THEN "30 We feel we are falling behind technologically." IF "30 We feel we are falling behind technologically" AND "40 We purchase and implement several AMTs," THEN "50 We may choose an AMT that is not suited to our workers' skills." IF "50 We may choose an

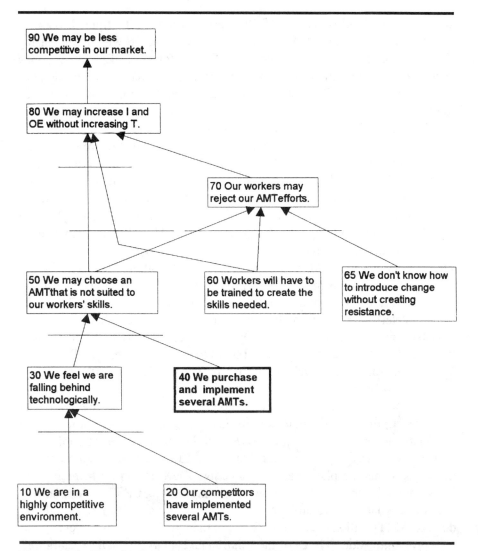

Figure 11.6   Negative Branch Reservation for "We May Select an AMT that Is Not Suited to Our Workers' Skills"

AMT that is not suited to our workers' skills" AND "60 Workers will have to be trained to create the skills needed" AND "65 We don't know how to introduce change without creating resistance," THEN "70 Our workers may reject our AMT efforts." IF "50 We may choose an AMT that is not suited to our workers' skills" AND "60 Workers will have to be trained to create the skills needed," THEN "80 We may increase I and OE without increasing T." IF "70 Our workers may reject our AMT efforts," THEN "80 We may increase I and OE without increasing T." IF "80 We may increase I and OE without increasing T," THEN "90 We may be less competitive in our market."

The third NBR—"Other innovations may be more productive"—is provided in Figure 11.7. Again, the base is the same. IF "10 We are in a highly competitive environment" AND "20 Our competitors have implemented several AMTs," THEN "30 We feel we are falling behind technologically." IF "30 We feel we are falling behind technologically" AND "25 We don't know how to choose AMTs to increase organizational productivity" AND "40 We purchase and implement several AMTs," THEN "50 We may choose the wrong AMTs to improve organizational productivity." IF "25 We don't know how to choose AMTs to improve organizational productivity," THEN "70 Other innovations may be more productive." IF "50 We may choose the wrong AMTs to improve organizational productivity" AND "60 The weakest link determines the strength of the chain," THEN "70 Other innovations may be more productive." IF "70 Other innovations may be more productive," THEN "80 We may increase I and OE without increasing T." IF "80 We may increase I and OE without increasing T," THEN "90 We may be less competitive in our market."

If we combined all of these trees into one, it would be more effective as a graphic argument and would help us determine what actions are needed to eliminate the negative consequences. Recall the initial conversation Bill had with his supervisor, who thought that the AMTs might solve some operational problems the plant was experiencing. The first step in trimming the NBR might be to take the action of constructing a CRT of the plant to determine what the core problem(s) is before we go any further. It might be that the AMTs will solve the problem, but they probably won't. Identify the core problem. Address it. Does the solution take a large capital expenditure? Is the constraint in the plant? What does the customer need? Better quality? Shorter lead times? More variety? Will the AMT address this specific customer need or will it merely reduce head count? Get the focus right! Will the solution increase throughput? Remember to shoot with a rifle instead of a shot gun. A small capital investment may increase organizational productivity significantly. Also, Bill may not have to spend any money to realize a significant improvement. Certainly Bill needs to approach this situation with his boss carefully. Bill's first action may be to get Bob, the CEO, to read

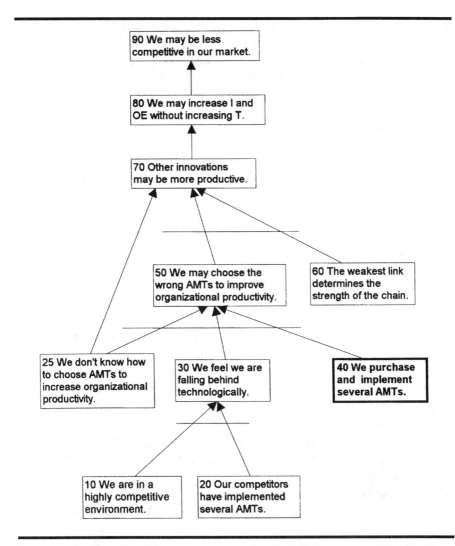

Figure 11.7   Negative Branch Reservation for "Other Innovations May Be More Productive"

Chapter 4 of *The Goal*,[4] then the whole book, next *Reengineering Performance Measurements*,[5] and then *It's Not Luck*.[1]

The NBR illustrates the causal relationships between an action and its negative effects. The objective is to specifically identify the linkages and determine where best to trim the negative effects. If Bob or Bill is successful in identifying an action to trim the negative effects, Bill can realize the positive effects from his CEO's willingness to invest in the organization without the negative effects of the investment being wasted, the workers being alienated, and the organization being less competitive.

SUMMARY

Goldratt's Thinking Processes may be the most important management tools developed this century. They certainly have been very useful in several independent and vastly different situations. There have been hundreds of successful implementations in manufacturing. The U.S. Air Force has used the Thinking Processes in its logistics and medical environment. In 1989 and 1990, Lieutenant Colonel Richard I. Moore made dozens of presentations and taught several workshops at the Air Force Institute of Technology to introduce Air Force personnel to the TOC logistics, measurement, and Thinking Processes paradigms. Moore, Eli Goldratt, John Blackstone, and Jim Cox taught the first military Jonah courses in 1990. Brigadier General Patricia Hinnenburg[2] was instrumental in moving the logistics implementation forward. Constraints management provides the foundation for the Air Force lean logistics system. Lieutenant General Charles H. Roadman II[3] is a leader in advancing the body of knowledge in using the Thinking Processes in a nonprofit and medical services environment. Ms. Kathy Seurken[4] has been just as impressive in advancing the use of the Thinking Processes in primary education, both in the classroom and in counseling. Several independent consultants have made tremendous productivity gains for organizations in the service sector, from grocery chains to car dealerships. The list of successes is continually growing in industries and applications.

Eliyahu M. Goldratt has been and is the primary developer of the Theory of Constraints. He is:

- The primary developer of the OPT software, developed in the late 1970s
- The author of *The Goal,* a business education novel that created a paradigm shift in the way many practitioners and academics view manufacturing
- A major critic of managerial and cost accounting
- A proponent of using the Socratic method of teaching
- A moving force in changing the face of manufacturing

- A creative problem solver
- The developer of a set of logical thinking tools—possibly his most important contribution

Our goal is for your organization to use this handbook successfully to achieve world-class status and continually improve its performance. In this manner, all of us can prosper.

REFERENCES

1. Goldratt, E.M., *It's Not Luck,* Croton-on-Hudson, NY: North River Press, 1994.
2. Goldratt, E.M., *What Is This Thing Called Theory of Constraints and How Should It Be Implemented?* Croton-on-Hudson, NY: North River Press, 1990.
3. Goldratt, E.M., " The Theory of Constraints in Industry." *APICS Constraint Management Symposium Proceedings,* Detroit, MI: American Production and Inventory Control Society, 1996.
4. Goldratt, E.M. and J. Cox, *The Goal: A Process of Continuous Improvement,* 2nd edition. Croton-on-Hudson, NY: North River Press, 1994.
5. Lockamy, A. and J.F. Cox, *Reengineering Performance Measurements: How to Align Systems to Improve Processes, Products, and Profits,* Burr Ridge, IL: Irwin Professional Publishing, 1994.
6. Hinnenburg, Patricia, "The Theory of Constraints and Lean Logistics." APICS Constraints Management Symposium Proceeding, Detroit, MI, April 1996, Falls Church, VA: APICS.
7. Roadman, C.H. II, "The Theory of Constraints and the United States Air Force Medical Service," APICS Constraints Management Symposium Proceedings, Phoenix, AZ, April 1995, Falls Church, VA: APICS.
8. Seurken, K., "TOC for Education." International Jonah Upgrade Workshop, Philadelphia, PA, September 1995, New Haven, CT: Avraham Y. Goldratt Institute.

INDEX

A

Academic articles, 132
Accounting, 23, 170
Administrative functions, 71
Advanced manufacturing technologies
     (AMTs), 300, 304
American manufacturing, 45
AMTs, see Advanced manufacturing
     technologies
A-plant
   characteristics of, 263
   product flow diagram for, 120
A-shape structure, 116
Assemble-to-order production, 174, 274
Assembly, 23, 263
   buffer, 94
   line, 54, 168, 169, 194
      capacity at, 217
      as constraints, 236
      stoppages, 192
      support of, 46
   operations, 204
   schedule, 84
      final, 94
      finite, 117
Assets, 56
A-structure, 31, 115
   characteristic of, 124
   control points used to manage, 125
   critical characteristics of, 231, 232, 233

example of, 131
plants, 229
   constraints for, 237
   drum-buffer-rope methods in, 227
primary control in, 117

B

Backward schedule, 84
Batch
   due dates, 278
   size, 291
Bill of material, 76, 102, 175, 202
Blockage, 105
Bottleneck, identifying, 58
BPR, see Business process reengineering
Breakage, 105
Budget, 241
   performance, 205
   preparation, 36
Buffer(s)
   analysis, region 1, 157
   appropriate size of, 90
   assembly, 94
   constraint, 98, 123
   definition of, 18
   effectiveness of, 195
   holes appearing in, 159
   informal assembly, 178
   lead time, 95
   management, 18, 73, 90, 192, 236